# Human Factors in Practice
## Concepts and Applications

# Human Factors in Practice
## Concepts and Applications

Edited by
### Haydee M. Cuevas
### Jonathan Velázquez
### Andrew R. Dattel

**CRC Press**
Taylor & Francis Group
Boca Raton London New York

CRC Press is an imprint of the
Taylor & Francis Group, an **informa** business

CRC Press
Taylor & Francis Group
6000 Broken Sound Parkway NW, Suite 300
Boca Raton, FL 33487-2742

International Standard Book Number-13: 978-1-4724-7515-2 (Hardback)

---

**Library of Congress Cataloging-in-Publication Data**

---

Names: Munoz Cuevas, Haydee, author. | Velazquez, Jonathan, author. |
Dattel, Andrew R., author.
Title: Human factors in practice : concepts and applications / Haydee M. Cuevas,
Jonathan Velazquez, Andrew Dattel.
Description: Boca Raton : Taylor & Francis, a CRC title, part of the Taylor & Francis
imprint, a member of the Taylor & Francis Group, the academic division of T&F
Informa, plc, [2017] | Includes bibliographical references.
Identifiers: LCCN 2017012378 | ISBN 9781472475152 (hardback : alk. paper) |
ISBN 9781315587370 (ebook)
Subjects: LCSH: Human engineering.
Classification: LCC T59.7 .M855 2017 | DDC 620.8/2--dc23
LC record available at https://lccn.loc.gov/2017012378

---

**Visit the Taylor & Francis Web site at**
**http://www.taylorandfrancis.com**

**and the CRC Press Web site at**
**http://www.crcpress.com**

*This book is dedicated to the many past pioneers and current leaders in the field of human factors. Their accomplishments have and will continue to inspire us to achieve greatness in our own research.*

# Contents

## SECTION I  Operator-Specific Considerations

## SECTION II  System and Environmental Considerations

## SECTION III   Putting Human Factors into Practice

# Foreword

I write this Foreword with a great deal of delight and humility. I am delighted to have the opportunity to introduce readers to a number of up-and-coming stars of our profession who write elegantly about our field. I am humbled by their willingness to ask me to write the Foreword.

Although a number of books providing an overview of human factors/ergonomics are available on the market, this one is unique—in two very different ways. First, one of the goals of this book is to highlight the work of early career professionals. The bright young minds who edited the book and authored the chapters were all within 10 years of earning their degrees when beginning this endeavor. This gives a voice to a fresh and more modern perspective on many topics within the field.

Second, this book is targeted at practitioners. That is, the book is not designed to be appealing to academics as an intellectual exercise, as many texts of this sort are. Rather, it is designed to illustrate how theories, principles, and concepts have been applied in a number of different domains. To that end, each chapter follows the same structure, with an introduction followed by fundamentals, methods, application of the methods, and future trends. In addition to the reference list, they also provide a list of key terms.

The book opens with a brief overview of human factors, followed by a description of our sensory modalities and our information processing capabilities. The book then describes methods for measuring human performance, which are key to understanding whether our designs will improve or hinder performance. For example, the chapter on situation awareness describes the challenges in using measurement techniques to understand the extent to which individuals are able to interpret the information they are receiving at any given time to assess the current situation and make decisions based on that assessment. The following chapter, on automation, also illustrates the difficulties in applying standard measurement techniques in deciding what level of automation could, or should, be used for a given system.

From there, the book goes on to explore the history of user-centered design, which initially developed out of work in the 1980s on human–computer interaction. This work, which was being done in psychology and computer science departments, morphed into the field currently known as user experience. The book then discusses the design of the workspace within which those humans and systems interact, with a focus on the individual, physical, psychosocial, and environmental factors that affect performance.

Within any application domain, it is important to be able to train users of the systems. The chapter on training design discusses what practitioners must know to develop, implement, and validate their organization's training programs. Finally, the book concludes with a look at a higher level of analysis, with a focus on sociotechnical systems, also known as macroergonomics.

Taken together, these chapters provide an introduction to many of the most important concepts in our field. I hope you enjoy reading it and find the guidance provided by the authors useful and applicable to your work.

**Deborah A. Boehm-Davis**
*Dean of the College of Humanities and Social Sciences*
*George Mason University*

# Editors

**Haydee M. Cuevas, PhD**, is an assistant professor in the College of Aviation, School of Graduate Studies at Embry-Riddle Aeronautical University, Daytona Beach, Florida. Prior to joining Embry-Riddle, she worked for over seven years as a research scientist at SA Technologies, Inc. She earned her PhD in applied experimental and human factors psychology from the University of Central Florida.

**Jonathan Velázquez, PhD**, is an associate professor and assessment coordinator for the Inter American University of Puerto Rico. He earned his PhD in aviation from Embry-Riddle Aeronautical University and a master's in educational arts in university teaching from the Inter American University of Puerto Rico.

**Andrew R. Dattel, PhD**, is an assistant professor in the College of Aviation, School of Graduate Studies at Embry-Riddle Aeronautical University in Daytona Beach, Florida, where he also is the director of the Cognitive Engineering Research in Transportation Systems (CERTS) Lab. He earned his PhD in human factors from Texas Tech University.

# Contributors

**Julian Abich, PhD**, is a senior human factors engineer at Quantum Improvements Consulting, Inc. He earned his PhD in modeling and simulation from the University of Central Florida.

**Debbie Ashmore, MS, MA**, is a senior human factors engineer at Lockheed Martin Mission Systems and Training in Moorestown, New Jersey, currently supporting Aegis Modernization Human Systems Integration and the EADGE-T program. During her time at Lockheed Martin, she has been applying human factors engineering on efforts for the U.S. Navy, U.S. Air Force, and U.S. Marines and several commercial aviation and maritime projects. She earned an MS in aeronautical science and human factors engineering from Embry-Riddle Aeronautical University and an MA in communications from Colorado State University.

**Ronald Boring, PhD**, is a human factors scientist at Idaho National Laboratory, where he has worked as principal investigator on research projects for the U.S. Nuclear Regulatory Commission, NASA, the U.S. Department of Energy, the Canadian Nuclear Safety Commission, the Joint Warfare Analysis Center, and the Norwegian Research Council.

**Deborah DiazGranados, PhD**, is an assistant professor in the School of Medicine of Virginia Commonwealth University. With more than 10 years of research and consulting experience, she has focused on teamwork, leadership, and team/leader effectiveness. She earned her PhD in industrial/organizational psychology from the University of Central Florida.

**Igor Dolgov, PhD**, is an associate professor of psychology at New Mexico State University, where he leads the Perception, Action, and Cognition in Mediated, Artificial, and Natural Environments (PACMANE) Laboratory. He earned his PhD in psychology (cognition, action, and perception) and arts, media, and engineering from Arizona State University.

**Raegan M. Hoeft, PhD**, is the director for Design Interactive's Federal Solutions division. Prior to joining Design Interactive, she spent six years working in the Human-Centered System Division at Lockheed Martin Mission Systems and Sensors and three years working at Electronic Ink, a design consulting firm. She earned a PhD in Applied Experimental and Human Factors Psychology from the University of Central Florida.

**Caroline Joseph, PhD**, is a professional engineer with a background in both mechanical and industrial engineering (ergonomics). She previously worked as a safety supervisor at Mortenson Canada Corporation and has experience in manufacturing, construction, and research settings. She earned a PhD in industrial engineering with a concentration in ergonomics from the University at Buffalo, The State University of New York.

**Elizabeth K. Kaltenbach, MA**, is a user research specialist with Sonos, Inc., with a background in human factors and engineering psychology. Her interests include acoustics, networked products, interaction and interface design, human–computer interaction, trust in automated systems, decision making, human performance, and video games.

**Joseph R. Keebler, PhD**, is an assistant professor of human factors and systems at Embry-Riddle Aeronautical University. He has over 10 years of experience in conducting experimental and applied research in human factors, with a specific focus on training and teamwork in military, medical, and consumer domains. He earned his PhD in applied experimental and human factors psychology from the University of Central Florida.

**Ahmed S. Khalaf, MA**, is currently a PhD student in computer science at New Mexico State University, working under the supervision of Dr. Zachary O. Toups. His research focuses on developing wearable interface designs to direct agent teams with adaptive autonomy. He also works as a lecturer at Al-Baha University, Saudi Arabia.

**Nathan Lau, PhD**, is assistant professor at the Grado Department of Industrial and Systems Engineering at Virginia Tech. He has published more than 30 research articles on interface design and human performance for the nuclear, medical, and petrochemical industries. He earned his PhD from the University of Toronto.

**Elizabeth H. Lazzara, PhD**, is an assistant professor of human factors and systems at Embry-Riddle Aeronautical University. She researches teamwork, team training, simulation-based training, performance assessment, and training evaluation in high-stakes domains. She earned her PhD in applied experimental and human factors psychology from the University of Central Florida.

**Rian Mehta, MA**, is working toward his PhD in aviation sciences with a specialization in human factors from the College of Aeronautics at the Florida Institute of Technology. He earned a master's in applied aviation safety. His publication research areas have focused on consumer perceptions and cockpit configurations.

**Dan Nathan-Roberts, PhD**, is an assistant professor in industrial and systems engineering at San José State University. His research is focused on human–computer interaction and ergonomics. He earned his PhD in industrial and operations engineering from the University of Michigan.

**Lauren Reinerman-Jones, PhD**, is the Director of Prodigy at the University of Central Florida's Institute for Simulation and Training, where she leads an interdisciplinary team of 35+ focusing on assessment for understanding, improving, and predicting human performance and system design. She earned her PhD in human factors psychology from the University of Cincinnati.

**Stephen Rice, PhD**, is an associate professor of human factors at Embry-Riddle Aeronautical University. His research interests include automation, trust, modeling, and consumer perceptions. He earned his PhD in aviation psychology from the University of Illinois at Urbana–Champaign.

**David Schuster, PhD**, is an assistant professor of psychology at San José State University. His research centers on understanding individual and shared cognition in complex environments. He earned his PhD in psychology specializing in applied experimental and human factors psychology from the University of Central Florida.

**Grace Teo, PhD**, is a faculty research associate at the University of Central Florida's Institute for Simulation and Training focusing on advancing assessment by connecting theory and application with particular attention to closed-loop human–robot teaming. She earned her PhD in human factors psychology from the University of Central Florida.

**Zachary O. Toups, PhD**, is an assistant professor of computer science at New Mexico State University, directing the Play and Interactive Experiences for Learning (PIxL) Lab. He earned his PhD in computer science from Texas A&M University.

**Shanqing Yin, PhD**, is a senior principal human factors specialist at KK Women's & Children's Hospital in Singapore. His current role focuses on the application of human factors principles and methodologies toward improving patient safety as well as clinical efficiency. He earned a BSc in psychology from the University of Illinois at Urbana–Champaign and a PhD in Human Factors & Systems Engineering from Nanyang Technological University, Singapore.

# 1 Editors' Introduction

*Haydee M. Cuevas, Jonathan Velázquez,*
*and Andrew R. Dattel*

Simply stated, the field of *human factors* facilitates the interactions between humans and technology. More formally, human factors, frequently used interchangeably with the term *ergonomics*, can be defined as "the scientific discipline concerned with the understanding of interactions among humans and other elements of a system, and the profession that applies theory, principles, data, and other methods to design in order to optimize human well-being and overall system performance" (International Ergonomics Association, 2017). Human factors traces its origins to World War II (WWII), when advanced instruments were being installed in complex aircraft and continues to be a critical component for promoting human performance in complex domains (for a brief retrospective, see Roscoe, 1997).

World War I brought a need for sophisticated technology and equipment. Although many advances during this period were made in aeromedical research, testing, and measurement, prior to WWII, the focus was on *designing the human to fit the machine* instead of *designing the machine to fit the human*. After WWII, technological advances focused more attention on the human element and designing for people's capabilities and limitations. The decades following WWII saw a prolific array of military-sponsored and academic research. This newfound interest in human factors brought forth advancements in our understanding of the factors influencing performance, ranging from human error and safety to team dynamics and organizational culture. Indeed, programs such as Crew Resource Management and Safety Management Systems both have their roots in human factors research (Velazquez & Bier, 2015).

Human factors practitioners analyze the factors (e.g., human information processing, situation awareness, mental models, workload and fatigue, human error) that influence decision making and apply this knowledge to identify potential hindrances to successful task performance, at both the individual and team level. They also evaluate how the design of advanced technology (e.g., automation, unmanned systems) can improve safety and performance but can also lead to unforeseen consequences, including changes in operator roles and responsibilities and the nature of their work.

Several books on human factors already exist in the literature (e.g., Helander, 2005; Lehto & Landry, 2012; Proctor & Van Zandt, 2008; Remington, Bochm-Davis, & Folk, 2012; Salvendy, 2012; Wickens, Gordon-Becker, Liu, & Lee, 2004). However, these in-depth volumes may be too advanced for the non–human factors reader and some are targeted at a specific discipline (e.g., aviation, engineering, human–computer interaction). In contrast, this edited book provides a succinct review of fundamental human factors concepts, presented at a level that can be easily understood by practitioners with no prior knowledge or formal education in human factors.

Thus, our book is targeted at practitioners rather than academics—a resource to be read at the workplace rather than used as a textbook in the classroom. The goal is to illustrate the authors' diverse perspectives on the application of human factors to address real-world problems across a variety of domains. The multidisciplinary background of the authors and the range of topics discussed are intended to increase the usefulness of this book to a wide audience of practitioners.

All chapters in this book address a common overarching theme (application of human factors theories, principles, and concepts to tackle real-world problems) and follow a similar structure (with the exception of the concluding chapter) to ensure consistency across chapters. The major sections in each chapter are listed in Table 1.1.

The chapters in this book are organized into two major parts. Section I focuses on operator-specific considerations and is comprised of four chapters: Senses in Action, Cognition in Action, Measuring Human Performance in the Field, and Situation Awareness in Sociotechnical Systems. Section II addresses system and environmental considerations and is comprised of four chapters: Automation in Sociotechnical Systems, User-Centered Design in Practice, Workspace Design, and Training Design. Our book concludes with Section III, a broader perspective on human factors in

## TABLE 1.1
## Chapter Section, Description, and Question Answered

| Section | Description | Question Answered |
|---|---|---|
| Introduction | Concise description of the topic's importance and relevance to the real world. | Why should I care? |
| Fundamentals | Discussion of relevant human factors concepts, theories, and principles. | What do I need to know? |
| Methods | Description of relevant human factors methods as well as guidance on useful resources to obtain more detailed information. | What tools can I use? |
| Application | Using a case study approach, this section will succinctly demonstrate the application of the human factors concepts presented in the Fundamentals section to address the current real-world problems described in the Introduction. | How can human factors solve some of these real-world problems? |
| Future Trends | Implications for future research in this area as well as new domains to be explored. | What is trending/what is next in this area? |
| Conclusion | Succinct wrap-up of the chapter. | What is the take-home message? |
| References | List of all references cited. | Who/what are your sources? |
| Key Terms | List of 6–10 key terms and definitions to enhance the reader's understanding of the chapter's content. | What human factors concepts did I learn? |

sociotechnical systems. Two appendices at the end of Chapter 10 provide useful tips and supplemental resources for the human factors practitioner.

Human factors remains a multidisciplinary profession and is likely to spread to new occupations as complex technology becomes more ubiquitous across domains. Today, individuals from a variety of fields such as psychology, engineering, and computer science can apply these skills to improve how people interact with systems and services. We hope you will find this book useful by providing practical information about human factors concepts relevant to the types of real-world problems you encounter in your work.

## REFERENCES

Helander, M. (2005). *A guide to human factors and ergonomics* (2nd ed.). Boca Raton, FL: CRC Press-Taylor & Francis Group. ISBN 9780415282482.

International Ergonomics Association (IEA). (2017). What is ergonomics? Retrieved from http://www.iea.cc/whats/index.html

Lehto, M. R., & Landry, S. J. (2012). *Introduction to human factors and ergonomics for engineers* (2nd ed.). Boca Raton, FL: CRC Press-Taylor & Francis Group. ISBN 9781439853948.

Proctor, R. W., & Van Zandt, T. (2008). *Human factors in simple and complex systems* (2nd ed.). Boca Raton, FL: CRC Press-Taylor & Francis Group. ISBN: 978-0805841190.

Remington, R. W., Boehm-Davis, D. A., & Folk, C. L. (2012). *Introduction to humans in engineered systems*. Hoboken, NJ: John Wiley & Sons. ISBN: 978-0470548752.

Roscoe, S. N. (1997). The adolescence of engineering psychology. In S. M. Casey (Ed.). *Human factors history monograph series* (Vol. 1, pp. 1–9). Retrieved from http://www.hfes.org/Web/PubPages/adolescence.pdf

Salvendy, G. (Ed.) (2012). *Handbook of human factors and ergonomics* (4th ed.). Hoboken, NJ: John Wiley & Sons. ISBN: 978-0-470-52838-9.

Velazquez, J., & Bier, N. (2015). SMS and CRM: Parallels and opposites in their evolution. *Journal of Aviation/Aerospace Education & Research, 24*(2), 55–78.

Wickens, C. D., Gordon-Becker, S. E., Liu, Y., & Lee, J. D. (2004). *An introduction to human factors engineering* (2nd ed.). Upper Saddle River, NJ: Pearson Education. ISBN: 978-0131837362.

# Section I

*Operator-Specific Considerations*

# 2 Senses in Action

*Lauren Reinerman-Jones,
Julian Abich, and Grace Teo*

## CONTENTS

## INTRODUCTION

In one way or another, human factors specialists and researchers are either directly or indirectly assessing the human senses in order to determine the optimal ways of designing a system or device that will meet the limitations of those senses or help expand them beyond their natural state. For example, mobile devices are one of the most common forms of global technology. Many mobile device companies have leveraged the guidelines and principles from the human factors discipline to determine the physical size and shape of the device, how and where icons should appear on the device, the types of feedback presented to the user, and a multitude of other important questions that helps inform the most efficient, effective, and safe ways of interacting with the device. All of the answers to these questions require an understanding of how the human senses work, what their limitations are, and how to get the most out of them in terms of attention and information processing.

## FUNDAMENTALS

The traditional approach to understanding the human senses concentrated on five primary senses: sight, hearing, touch, smell, and taste. Extensive research has found that many more senses actually exist, and each has its own mechanisms through which we perceive the world and ourselves. Some senses focus on external stimuli, such as perceiving temperature, while others are geared to internal responses, such as hunger. Some would argue we have more than 20 distinct senses. The first part of this chapter, "What Are the Senses?" will first focus on a brief overview of the five basic senses and touch upon the many others that have been identified. Next, "Senses in Human Performance Theory" describes the theoretical contribution of the senses in regard to human information processing and performance. Following, the Method section takes a human factors approach to describing assessments of the senses in terms of human performance. Finally, the chapter presents a couple of examples illustrating human factors applications geared toward assessing the senses.

### WHAT ARE THE SENSES?

Aristotle is credited with traditional classification of the five sense organs. Humans have five main sense organs: eyes, ears, skin, nose, and tongue, which have evolved over time to preserve and protect the species. It is through receptors in these sense organs that stimuli are transformed to signals, either chemical, electrical, neurological, or a combination thereof, and transmitted to the brain for further processing. It is within the brain that meaning is associated with these signals. In other words, the information out in the world is quite meaningless until a human interprets it in a way that can be used in a meaningful way. The stimuli can usually be quantified objectively without human assistance, but it is in the brain that quality is associated with the signal. It is a phenomenon that occurs in most cases to preserve the species and assist with human evolution. For example, the concentration of sugar in a fruit juice can be measured by determining the ratio of sugar to water in a specified amount of juice extracted from a fruit, but determining whether it is sweet requires the ratio to exceed a threshold that a human will register as sweet after the concentration has triggered a response in the brain. The concentration is quantifiable, but the sweetness is assessed qualitatively.

Let us now turn to each of the five main human senses and discuss each individually. We are not going to delve into great detail about the anatomical structures or process through which stimuli are converted into brain signals since there are a plethora of textbooks that explain this more eloquently. Instead, we shall focus more on making sure we have a common understanding of the senses in order to discuss them from a human factors perspective. We will end this section with a quick reference to other body senses that have been identified over the past decades.

### Sight (Vision, Ophthalmoception)

Sight is the most dominant sense, with about 80% of cortical neurons responding to stimulation of the eyes. It refers to the ability to capture, perceive, identify, and discriminate images and colors through the eyes. When a stimulus in the environment is attended to, the image is projected through multiple layers and fluid in the eye onto the retina (the back lining of the eye that contains light-sensitive receptors and other neurons). The dominant neurons that line the retina consist of two photoreceptors: rods and cones. Rods are sensitive to light, do not distinguish colors, and are in higher abundance than cones (approximately 120 million). Cones are sensitive to color, are much more concentrated in an area called the macula (the center of which is called the fovea centralis and does not contain rods), and are more scarce than rods (approximately 6 to 7 million). Through transduction (conversion from one form of energy to another), the neurons that line the retina transform the external image into an electrical signal. This activation sets off a network of interconnected neurons that communicate with each other to send this electrical signal to the visual processing areas of the brain through the optic nerve (the head of which is called the optic disk and is considered the blind spot of the eye because it does not contain any photoreceptors).

### Hearing (Audition, Audioception)

Hearing refers to the ability to capture, perceive, identify, and discriminate sounds. Sounds waves are funneled in by the outer ear (pinna) and travel through the ear canal to the ear drum, where they are transformed into mechanical vibrations. These vibrations travel through the hearing bones (malleus, incus, and stapes) in the middle ear to the fluid-filled chambers of the inner ear (cochlea). The cochlea contains the sensory hearing cells, the primary two being the outer and inner hair cells. The outer hair cells soften or amplify sounds while the inner hair cells transfer sound information to the auditory nerve, which then transfers this information to the various regions of the brain that process sound.

### Touch (Somatosensation, Tactition, Mechanoreception, Kinesthesia, Tactioception)

Touch refers to the ability to capture, perceive, identify, and discriminate physical sensations. The skin contains about 5 million receptors and is the primary organ through which the sense of touch begins. As an external stimulus comes in contact with the skin (epidermis and dermis), mechanoreceptors (sensory neurons) transform mechanical stimulation (pressure, stretching, vibration) into an electrical signal that is transmitted through the peripheral nervous system to the central nervous system (brain and spinal cord), where a response to the stimulus is generated and sent back to the muscles through motor neurons. Four main mechanoreceptors are found in the skin: Merkel receptor, Meissner corpuscle, Ruffini cylinder, and Pacinian corpuscle.

Other mechanoreceptors called nociceptors are responsible for the sensation of pain and are more than just the overloading of sensors. Some respond to extreme temperatures (thermal nociceptors), chemical irritants (chemical nociceptors), or excessive pressure (mechanical nociceptors). Nociceptors are also found within the body (bones, joints, organs) and account for internal pain sensation.

Most touch sensory information travels to the brain, but at times, where an immediate response is needed to avoid injury, the spinal cord can send a motor signal to the muscles directly. This is referred to as the Spinal Reflex Arc, which is a reflexive response from the spinal cord to information received from the nociceptors that constitute an immediate response. These are generally triggered by nociceptors (pain sensors).

## Smell (Olfaction, Olfacoception)

Smell refers to the ability to capture, perceive, identify, and discriminate smells or scents. Smell is one of two chemical senses (the other is taste). Unlike sight, hearing, and touch, which all are results of nerve endings responding to stimuli, chemical senses are unique in that they require the body to take in molecules. These senses are sometimes referred to as "gatekeepers" because they protect the body by distinguishing among potential harmful substances. As odorant molecules pass into the nose, the molecules encounter the olfactory mucosa (where the olfactory sensory neurons and receptors are located). When an odorant molecule binds with a receptor (and there are over 350 different types), just like the other senses, the information is transformed into an electrical signal that activates the sensory neurons. The signal then reaches the olfactory bulb (which is part of the brain), where it accumulates signals from many neurons and then transmits them to the other olfactory processing areas within the brain. Again, it is within the brain that we are able to make sense of the signal to recognize and identify odors.

## Taste (Gustation, Gustaoception)

Taste refers to the ability to capture, perceive, identify, and discriminate flavors. This is the second chemical sense as described before when discussing olfaction. A substance first comes in contact with the tongue, which is covered with small bumps called papillae. These papillae are found on the tongue, palate, and epiglottis. It is within three of the four types of papillae that the taste buds lie (about 10,000 of them). Each taste bud contains about 50–100 taste cells, which have pores (receptor sites). When these sites are activated by chemical substances, a transformed electrical signal is generated and transmitted through many different nerves from the tongue to many areas in the brain, even some that share connections with the olfactory pathway. The taste buds allow us to detect sweet, sour, bitter, salty, and umami (a Japanese term that has no direct synonym in the English language but is described as "savory").

## Additional Senses

The five senses model is the one often taught in school, but many psychologists, philosophers, and physiologists purport many more. Some argue that humans have some 20+ senses and more yet to be discovered. Here are some commonly accepted additional senses.

- Proprioception: awareness of the body parts and in relation to each other
- Equilibrioception: Sense of balance. Vestibular system. Although this may not have much to do with hearing, it does have to do with a combination of the sight, inner ear, and various body senses (e.g., proprioception, tension sensors, etc.)
- Tension sensation: monitor muscle tension
- Stretch reception: sense dilation of blood vessels, often involved in headaches
- Chemoreception: the medulla detects blood-borne hormones and drugs, and is involved in vomit reflex
- Magnetoreception: Ability to detect magnetic fields. Fairly weak in humans, but do have some sense of magnetic fields
- Time: Debated because no single mechanism found that allows humans to perceive time. What little we know about this makes it difficult to assess how perception of time affects human performance, especially during high stress and long duration tasks

## SENSES IN HUMAN PERFORMANCE THEORY

Immanuel Kant: Our knowledge of the outside world is dependent on our modes of perception. Sensation and perception are interrelated processes that work together to allow humans to extract or append meaning to external and internal stimuli (Goldstein, 2007). Sensation refers to the accumulation and processing of information from stimuli (physical and imaginative) that impinge on the sense receptors. This information is sent to the brain for perceptual processing to take place, but the sensation alone cannot effectively drive human performance. It is the perception of the sensation that contributes to the meaning and action of human behavior.

Perception refers to the conscious experience generated from the interaction among the processed sensation and the brain's neural network. It is the way we interpret the sensations and gives way for recognition to occur. If a fire alarm is triggered, the raw sound that is heard does not mean anything except it is a loud noise that may be bothersome. The perceptual process is where the meaning of the sound is connected, such as "a fire is present and I need to leave the building." This meaning can be gathered from information from the signal itself and is built up from the components of the signal or other related signals, such as people running toward an exit. This is referred to as bottom–up processing. Conversely, if the signal sent from the sensory receptors is diminished, then previous experience, knowledge, or expectations may help determine what the signal means. This is referred to as top–down processing. Both are important aspects of human information processing that affect human performance. Taking into account the concepts of sensation and perception, some human factors theories and models have been proposed to designate their roles in supporting human performance.

Many human factors interaction models have been proposed, each having their merits, but the essential components in all of them involve a closed-loop system between a human and a system or machine. Information is processed through the human sensory systems which inform the appropriate action to be taken. The human then takes action on a system, the system processes that action and displays the results, closing the loop and providing the human with sensory information and thus the cycle repeats until a goal is achieved. These models clearly illustrate the importance of the human sensory system to initiate or facilitate this sequence of interactions. Without acquisition of information from the environment (or internally), cognitive processing would literally be senseless. The following section will focus on describing some human performance models and theories and the role the human sensory systems play to support efficient, safe, and successful interactions.

## Theory of Information Processing

The Human Information Processing model (for full description, see Wickens & Hollands, 2000) integrates commonly agreed upon structures for information processing gathered from extensive research and it provides a useful framework for performance prediction. This closed-loop model does not designate a specified starting point because information processing could occur at any state, such as in response to an environmental stimulus or from voluntary activation. For the purposes here, we will focus on information processing with respect to sensory stimulation.

As described in the previous section covering the human senses, the receptors capture and transform stimulus information into a message (neural activation) that is processed through sensory pathways to processing centers in the body where perception occurs. Each sense has an associated storage capacity and duration (termed short-term memory store) that indicates how long information could be retained by the sensory receptors. The allocation of attentional resources to the signals generated by the sensory receptors determines which information is further processed, but the quality of information to be transmitted greatly depends on the receptor's capabilities to capture stimulus information, which in turn affects the level of attention allocated to a stimulus.

## Theories of Attention and Mental Resources

Humans have limited attentional capacity and resources. It is self-evident by the inability to attend to every single conversation in a crowded area. It is therefore important for a human to attend to the most important information presented in the environment, but this poses a challenge as well because attention to more than one source of information may be necessary or a salient signal may be distracting.

Resource theory suggests that humans have a single-pool of undifferentiated resources available to support information processing and humans are limited in their capacity to process information (Kahneman, 1973). However, there is flexibility in the ways that these resources are distributed to meet the demands of a task. The Multiple Resources Theory takes this a step further and suggests that there is a differentiation among resource pools and those resources can be distributed separately to tasks requiring different cognitive processing (Wickens, 2002), hence accounting for performance of many time-sharing tasks (Wickens & Hollands, 2000). There is evidence to support both theories, but what is important when discussing the human senses is that there is a limitation to how much information can be attended to at any given time.

Attention can be thought of in three ways: selective, focused, and divided. Selective attention refers to the intentional choice to process specific information, but the chosen information may not always be the most optimal for the task. Focused attention refers to the decision to ignore information that is distracting and to concentrate on the most important information source. Divided attention refers to focusing attention on more than one source of information, sometimes intentional but occasionally not so much. As mentioned before, in regard to the human senses, the quality of the signal processed by the sensory receptors will affect the level of attention. Designers understand these attentional characteristics and exploit these resources using various cues to either enhance the likelihood that a sensory signal will be processed or reduce the possibility of inadvertently processing unwanted information.

## Signal Detection Theory

When discussing sensory information processing, there are two main states that exist, either the signal is present or it is not. Signal detection theory provides a framework to classify the decisions made regarding the presence of a signal, in this case a signal presented from sensory receptors (Green & Swets, 1966). Combining the two discrete states of the signal with two response states creates a matrix that defines all possible outcomes of detecting a signal: hit, miss, false alarm, or correct rejection. Once sensory receptors are activated by a stimulus, the extent to which that signal is salient enough for detection will determine whether that signal is attended to or not. It is assumed that noise is always present and a signal can be detected if there is enough evidence to distinguish it from the noise. The cocktail phenomenon is a perfect example of focusing attention on an auditory stimulus while being inundated by other auditory information. Imagine being at a party and taking part in a conversation with a group of people. It is quite possible to focus attention solely on that group of people. Then suddenly across the room of other loud conversations someone says something that is very familiar to you, such as your name, and you cannot help but attend to that even though you were focused and potentially speaking to someone else. The combination of previous knowledge, context, and signal quality enhanced the likelihood of detecting a signal that was present amongst the noise of the crowd.

Many complex tasks require performers to detect signals that are hidden among noise, such as an airport baggage screener trying to identify a weapon in a bag among the clutter of clothes and toiletries. It is for these reasons that certain items, such as laptops, need to be removed from your bag when going through airport security and sometimes the screeners need to check your bag further because the x-ray image was unclear for them to scan the items.

## Change Detection

At times, signals may be very salient, yet they go undetected. This is especially true when trying to detect changes. Change detection is the processing involved in noticing a change, identifying the type of change, and locating the change (Rensink, 2002). Each sensory system has a different threshold at which a change can be detected (this will be discussed in more detail in the Methods section). Further, it may seem rational to believe that a change occurring to the same sensory system should be detected more often, but this is not always the case. The inability to notice change is referred to as change blindness (Simons & Levin, 1997). A very famous study required participants to view a visual scene of people passing a ball around and were asked to count how many passes occurred (Simons & Chabris, 1999). While this was occurring, a person dressed as a gorilla walks through the center of the group. When participants are asked what animal walked through the scene, most are unable to provide an answer. When the video is presented and the attention is focused on finding the animal, participants find it difficult to believe that they were unable to detect it the first time around. This emphasizes the importance of attention in detecting signals.

A human factors solution to ensure changes or signals are detected is to provide redundant signals, meaning the presentation of the same signal but to different senses. Fire alarms are a perfect example because they not only provide a loud and intentionally obnoxious auditory alert, but also there are usually bright blinking lights associated with the sound to ensure the message is received.

## Grouping Principles

Gestalt theorists believe that the whole is greater than the sum of its parts such that the perception of a group of stimuli can be very different from the perception of each element in the group. The ways in which these stimuli are grouped by the brain have been extensively investigated resulting in a list of grouping principles or laws. These principles are well established and can be extensively described in many sensation and perception textbooks, so a brief description here will suffice. These include the following:

- Similarity: things that look alike tend to belong together
- Proximity: things that are close to each other belong together
- Continuity: the preference for seeing stimuli as connected or continuous
- Closure: the tendency to complete shapes that are not complete
- Symmetry: a tendency to see objects as balanced around a center form
- Common fate: the tendency to see things moving together as a group

Human factors designers leverage these natural tendencies to perceive stimuli in such way to predict human performance behavior and thus evoking an expected response from their designs. It requires skill and practice to effectively implement these principles, and simply adding them to a design does not guarantee the expected behavioral outcome will be achieved.

## METHODS

As described above, sensation and perception are key processes that must occur in order to achieve a level of behavior. Human factors designers want to ensure that a specific response behavior is elicited by their designs, otherwise, they must be reevaluated and, ultimately, redesigned. This process requires the assessment of the designs and the effects they have on the sensation and perceptual processing systems. Borrowing from the psychometrics and psychophysiological literature, we will discuss some subjective and objective approaches to measuring the senses.

### Subjective

Perceiving colors and taste are phenomena that exist as the result of an interaction between the sense receptors and the processing areas in the brain. Color does not exist unless a human is there to transform the light frequencies into distinct bands with individual identity. Sound only has pitch if a human is there to transform vibrations of air molecules into something meaningful. Therefore, it would seem intuitive to assess the human senses through subjective means.

Threshold assessment is used to determine when a stimulus causes a change or response in the observer. There are two major categories of threshold assessment: absolute and difference. Absolute threshold is the point at which the stimulus triggers a response from the sense receptors. In other words, it is the least amount of information needed by your senses for it to be detected. The following are various type of thresholds considered in the measurement of sensation:

- Detection threshold: the lowest concentration of a smell or taste that can be detected.
- Recognition threshold: the lowest concentration at which quality of the smell or taste can be associated.
- Difference threshold: the ability to detect a change in the attended stimulus.
  - Just noticeable difference (Weber's Law) is the amount of change needed to detect that difference.

Human factors approach the sense of touch through the science of haptics. Haptics, the application of devices that leverage the touch sensory system, is divided into two main categories: tactile and kinesthetic. Tactile perception is the things you feel from the sensors in the tissue of your skin, such as your fingers. Kinesthetic perception is the things you feel from the muscles, joints, and tendons and help determine where the body is in relation to other body parts. Most commonly, tactile perception is assessed by human factors designers. Tactile acuity has been traditionally assessed using a two-point threshold approach, which assesses the minimum distance between two points that must be on the skin when stimulated to be perceived as separate. Grating acuity is done using a groove threshold approach which is a two-part test that first assesses the ability to detect the orientation (vertical or horizontal) of the grooves when pressed on the skin. Acuity is determined by the minimum spacing between the grooves that can be detected for that orientation.

Many specific approaches have been developed to identify points at which a sensation is noticed or recognized:

- Method of limits: Presentation of stimuli intensity in a step-wise ascending or descending order, starting each trial at a different intensity to find an average threshold.
- Method of adjustments: This approach is similar to the method of limits, but continuous stimulus intensity changes are used rather than step-wise changes. The least accurate approach but fastest method.
- Method of constant stimuli: Multiple trials of randomized stimuli intensities are presented. The intensity that is detected on 50% of the trials is considered the threshold. This method is the most accurate.

Interestingly, humans are able to distinguish between different odors but are not very good at identifying the source of them, although practice and feedback improve that dramatically.

Related to threshold assessment is magnitude estimation. Magnitude estimation is when a stimulus is compared to a "standard" stimulus and an observer determines the intensity difference of the stimulus by assigning a value that is proportional to their observation. Magnitude estimation could also occur with no standard stimuli. The main difference between threshold assessment and magnitude estimation is that threshold assessment simply identifies when a sensation is achieved, while magnitude estimation tries to determine how big of a difference is perceived between two sensations. There are some challenges to performing many of these assessments, but one main issue is that the sensory system does not continuously send out signals to the brain, regardless if the stimulus is still present.

Sensory adaptation is when the sense receptors cease responding to a stimulus that it is constantly exposed to for an extended period. This is an important concept to consider within the human factors field. If your body constantly responded to all of the senses it was exposed to, the information would be overwhelming to the point that it would inhibit your behavior. On the other hand, if there is certain important information that is constantly presented, then there could be a chance that information is missed, which can lead to potential catastrophic events. The limitation of these subjective approaches can be overcome with approaches that are more objective.

## OBJECTIVE

Effective, efficient, and safe human performance is ultimately the result of good design. Hence, assessing human performance can provide insight to the way the senses gather (or fail to gather) information from the environment to make decisions and drive actions. Applicable to all senses, time measures such as reaction and response time can provide practical information regarding how long it takes for a specific behavior to be elicited once a stimulus is generated.

Some measures are specific to some of the senses. Visual acuity is the ability to discern specific objects at a given distance. The Snell charts used at the office of an optometrist are used to assess visual acuity. Eye trackers are instruments used to capture how the eye moves when the person is viewing something. The metrics that are gathered from such a device include the number of fixations, fixation duration, gaze detection, number of saccades, pursuit, and the index of cognitive activity (ICA). All metrics, besides the ICA, are direct assessments of eye behavior that help determine how humans interact with visual stimulus. The ICA infers a level of mental effort based on the behavior of the pupil and has been correlated with perceived workload.

Most auditory perception is done through subjective means, except when using an optoacoustic emissions test. This test is used to stimulate the inner ear without the need of human perception. By sending in sound vibrations, the inner ear responds and sends back an almost inaudible sound that can be detected with the audio probe. This provides an objective measure to determine the hearing ability of an individual.

The use of neurophysiological sensing devices makes objective assessment of the senses possible in regard to capturing the effects stimuli have on sensory receptors when they are registered by the body. Electroencephalograms (EEGs) capture neuronal electrical activity in the cerebral cortex. There are pathways in the brain that are associated with processing information sent from sensory neurons and when these sensations are processed by areas of the cortex, the EEG can measure the extent to which that information is being processed. Along those same lines, the functional near-infrared spectroscopy (fNIRS) and transcranial Doppler (TCD) ultrasonography both assess metabolic functioning within the brain. Therefore, when sensation information is processed in the brain, activation of neurons must occur for that sense to be perceived and that activation requires "fuel" or sources of energy. The fNIRS and TCD assess the consumption and transfer of the brain's fuel, which is blood and oxygen, and helps human factors designers determine whether a specific design registered a response from the user.

Further, another common measure used by human factors practitioners is electrocardiogram (ECG). ECG assesses cardiac (heart) activity. This activity has been used to assess many human-related states, such as stress and relaxation. This measure appeals to human factors designers because of cost, application, and quality of information gathered. When a new display or interface design is implemented, ECG can help determine the type of response elicited in the human while interacting. For instance, the fire alarm that is perceived both auditorily and visually is designed to capture attention and increase alertness. When testing various auditory tones and visual cues, ECG can assess whether a specific physiological response is activated.

## APPLICATION

This next section describes two examples of the human factors application of sensory assessment to support interface and training design. There are endless examples, but we have chosen some that together encapsulate many of the previously discussed content.

## HOLOGRAPHS DISPLAYS

Display technology is transitioning from flat, two-dimensional (2D) displays to displays that provide three-dimensional (3D) perspective views with gross applications within many domains such as medical, military, educational, air-traffic control, and, of course, entertainment. Dynamic, interactive forms of content display have been shown to significantly increase learning performance over static display forms (i.e., textbook). Through the use of virtual reality, augmented reality, stereoscopic displays, and other technological approaches, 3D environments are readily and easily achieved. The limitation is that such displays usually require some type of viewing device, such as a headset or glasses, to see the 3D items. Additionally, these devices usually require some level of power supply, making portable systems less available or battery life an issue. Therefore, an interest in reduced technologically dependent 3D displays has led to a resurgence of printed holographic imaging. Holographic displays offer the advantage of portability, simultaneous viewing by multiple users, and are less expensive than physical models. Prior to the implementation of these types of displays, an understanding of the effects on human information processing and performance must be established.

Displays that reflect ecological similarities to real world perspectives may support visual sensory processing by reducing the mental effort to reconstruct 3D images from 2D ones (Smallman, St. John, Oonk, & Cowen, 2001). The reduced cognitive effort to process visual information from 3D displays theoretically implies that more mental resources are available to process other pertinent content information that is associated with the displayed information. Research is beginning to determine which types of information are best conveyed through 3D displays, such as conceptual or procedural information, in order to support a positive transfer of training to practical problems.

## HUMAN–ROBOT TEAMING

One major program the U.S. Army has undertaken is to advance the state of autonomous systems to support dismounted Soldier–Robot Teams. Dismounted refers to soldiers that are on foot, not in a vehicle, such as the infantry or Marines. The traditional approach to using a robot within this context requires a specified individual to teleoperate (manually control) a robot, while constantly attending to a visual display. This requirement poses a tremendous problem for dismounted soldiers because

a. the robot will only work if someone is moving it with a joystick or controls;
b. a specific person needs to be trained to control the robot;
c. the person controlling the robot can easily lose awareness of the surroundings; and
d. the operator is unable to hold a weapon for protection.

For these reasons, it is imperative that the state of robotic assets begins to reflect the same behavior as a human soldier teammate. In order for this progression to take place, the same forms of communication that occur between human soldier teammates should exist between soldiers and robots. One approach has focused on implemented multimodal communication (MMC) (Lackey, Barber, Reinerman-Jones, Badler, & Hudson, 2011). MMC is the conveyance of information through more than one sensory system (Partan & Marler, 2005). Messages can be conveyed that are redundant (the same signal but to different senses) or nonredundant (different signals are sent to different sensory systems simultaneously). Both are important approaches and have multiple effects on the user. The approach taken here focuses on redundant signaling to ensure the message is received during bidirectional communication with a robot teammate (Abich, Barber, & Reinerman-Jones, 2015; Barber et al., 2015; Barber, Lackey, Reinerman-Jones, & Hudson, 2013).

Understanding the environmental context is important when designing for soldiers. Dismounted soldiers are inundated with sensory stimulation, such as gunfire, loud crowded areas, weather, etc., and therefore, if critical information needs to be received by a soldier, then sending that information in more than one form is essential for that signal to be detected. The ways in which soldiers provide other human soldiers with information is usually done visually (hand gestures or body postures), auditorily (voice commands), and/or tactilely (tapping a soldier on the shoulder). The ability of the soldiers to process this information and use it to make decisions in the battlefield is important for the soldiers' safety and mission success. Therefore, understanding the availability of the soldiers' sensory system to process information is important when designing interactive approaches (visual, audio, or tactile displays) that will be used to communicate information to a soldier from a robot teammate.

## FUTURE TRENDS

With the current trajectory of technological growth, it can be anticipated that future advancements would involve further integration of human and technological capabilities and the direct enhancement of human senses. Innovations in computer science and robotics are likely to result in smarter technology that can "sense" humans (i.e., sensors) and adjust to them in return in much the same way as humans sense and adjust to each other. Such technological sensing would allow displays, controls, and interfaces to be dynamically adapted to the human across different tasks and situations.

Finally, human sensory capabilities may also be enhanced. Enhancements can be intranormal, where improvements are within the normal human range, or supernormal, where improvements extend beyond the normal human range (e.g., being able to hear out-of-range frequencies) and may involve acquisition of new traits (Brey, 2008). Scientists in this area have already shown the possibility of teaching rats to "see" infrared light, which is usually invisible to rats (Moyer, 2013). Nevertheless, this line of research has ethical implications on self-concept and species identity (Brey, 2008) and can be somewhat controversial.

## CONCLUSION

This chapter showed how human performance and safety can be understood in terms of how our sense organs function and how we process sensory information. It also described the different methods of measuring the senses and introduces the main concepts in measurement of the various senses. The examples cited illustrate how our understanding of our senses and perception guide the development of cutting-edge technology that can aid us in our work. The examples also show that since our senses are the means by which we acquire information, our understanding of the senses would not be adequate without the consideration of the environment in which we operate.

## REFERENCES

Abich, J. I., Barber, D., & Reinerman-Jones, L. (2015). Experimental environments for dis-mounted human–robot multimodal communications. In R. Shumaker & S. Lackey (Eds.), *Virtual, augmented and mixed reality: 7th International Conference, VAMR 2015, held as part of HCI International 2015* (Vol. 9179, pp. 165–173). Los Angeles: Springer International Publishing.

Barber, D., Abich, J. I., Phillips, E., Talone, A., Jentsch, F., & Hill, S. (2015). Field assessment of multimodal communication for dismounted human–robot teams. *Proceedings of the 59th Human Factors and Ergonomics Society 2014.* Los Angeles, CA.

Barber, D., Lackey, S., Reinerman-Jones, L., & Hudson, I. (2013). Visual and tactile interfaces for bi-directional human–robot communication. *SPIE Defense, Security, and Sensing. 8741*, pp. 1–11. International Society for Optics and Photonics.

Brey, P. (2008). Human enhancement and personal identity. In J. Berg Olsen, E. Selinger, & S. Riis (Eds.), *New Waves in Philosophy of Technology*. New Waves in Philosophy Series (pp. 169–185). New York: Palgrave Macmillan.

Goldstein, E. (2007). *Sensation and perception* (7th ed.). Belmont, CA: Thomson Wadsworth.

Green, D., & Swets, J. (1966). *Signal detection theory and psychophysics*. New York: Wiley.

Kahneman, D. (1973). *Attention and effort*. Englewood Cliffs, NJ: Prentice-Hall.

Lackey, S. J., Barber, D. J., Reinerman-Jones, L., Badler, N., & Hudson, I. (2011). Defining next-generation multi-modal communication in human–robot interaction. *Human Factors and Ergonomics Society Conference*. Las Vegas, NV: HFES.

Moyer, M. W. (2013). Brain implants could enhance our senses. *Scientific American.* Retrieved from https://www.scientificamerican.com/article/brain-implant-could-enhance-our-senses/

Partan, S., & Marler, P. (2005). Issues in the classification of multimodal communication sig-nals. *The American Naturalist, 166*(2), 231–245.

Rensink, R. (2002). Change detection. *Annual Review of Psychology, 53*(1), 245–277.

Simons, D., & Chabris, C. (1999). Gorillas in our midst: Sustained inattentional blindness for dynamic events. *Perception, 28*(9), 1059–1074.

Simons, D., & Levin, T. (1997). Change blindness. *Trends in Cognitive Sciences, 1*(7), 261–267.

Smallman, H., St. John, M., Oonk, H., & Cowen, M. (2001). Information availability in 2D and 3D displays. *IEEE Computer Graphics and Applications, 21*(5), 51–57.

Wickens, C. (2002). Multiple resources and performance prediction. *Theoretical Issues in Ergonomics Science, 3*(2), 159–177.

Wickens, C. D., & Hollands, J. G. (2000). Attention, time-sharing, and workload. In C. D. Wickens, J. G. Hollands, N. Roberts, & B. Webber (Eds.), *Engineering psychology and human performance* (Vol. 3, pp. 439–479). Upper Saddle River, NJ: Prentice-Hall Inc.

## KEY TERMS

**attention:** a cognitive process of selecting only parts of the perceivable world to focus on.

**change blindness:** inability to notice an obvious visual change.

**just noticeable difference:** amount of change in a stimulus needed to detect a difference.

**magnitude estimation:** method of measuring sensation that involves determining how big of a difference is perceived between two sensations.

**perception:** process of interpreting sensory information.

**sensation:** process of obtaining information through the senses.

**senses:** the means by which humans acquire information about an external or internal stimulus.

**sensory adaptation:** occurs when the sense receptors cease responding to a stimulus that it is constantly exposed to for an extended period.

**threshold assessment:** method of measuring sensation that involves identifying when a sensation is achieved.

# 3 Cognition in Action

*Shanqing Yin*

## CONTENTS

## INTRODUCTION

Practically everything we do requires us to think and process information. When we drive, we perceive the situation around us before reacting via the steering wheel. We figure out when and how would be a good route to take. We occasionally recall the wrong information or derive the wrong answer while doing mental calculations. Doctors attempt to make sense of all the patient's symptoms before coming up with a diagnosis and prescribing an appropriate intervention. Any task that involves a human operator would involve some form of cognition.

Like anything else relating to humans, our cognition is not perfect. Our cognitive activities involve multiple components interacting with one another as information is being processed in the mind. Driving a car involves attending to the traffic situation, processing the spatial information, and deciding on an appropriate output. While we generally perform this process efficiently and effortlessly, we can still fumble. Our driving performance deteriorates when we attend to a text while driving at the same time. We take a longer time to orientate ourselves when driving in an unfamiliar location. Despite noticing the sudden appearance of a deer, we may still not have enough time to react without incurring damage. There is a limit to our cognitive abilities, and being familiar with our cognition in action would help us mitigate our vulnerabilities and improve our performance.

We can understand our cognition by comparing it to an information processor, which is a system of receiving information in one form and transforming it into another. It typically involves four key stages: input, processing, storage, and output. A straightforward analogy would be our personal computers and peripheral accessories (Figure 3.1). Information is *input* into the computer through the keyboard, mouse, webcam, or even barcode scanner. The computer then *processes* the data while *storing* it in the random access memory (RAM) or the read-only memory (ROM) hard drives, eventually generating an *output* such as an image on the monitor

**FIGURE 3.1**    This typical computer setup is an example of an information processing system.

screen, printed document, or sounds from the speakers. Similarly, we attend to information, mentally process it in our minds, store it for future recall, and initiate a cognitive output such as making a choice or decision.

By using the information processing model, this chapter offers a digest on the core concepts surrounding cognition in human factors. The information processing model aids in identifying where cognitive performance issues are in tasks and whether proposed interventions are appropriate. An issue in which an important alert goes unattended would require increased saliency, whereas a complicated mental calculation can be supported through using computational tools. Conversely, a failure to appreciate this intricate cognitive process can result in wrong root-cause diagnosis and ineffective solutions. The following selection of books provides a more extensive exploration of human cognition and its relevance to human performance: *The Model Human Processor* (Card, Moran, & Newall, 1986), *Information Processing* (Wickens & Carswell, 2006), and *Engineering Psychology & Human Performance* (Wickens, Hollands, Banbury, & Parasuraman, 2015).

## FUNDAMENTALS

Components of our cognitive process can be specifically mapped out using the information processing model (Figure 3.2). We perceive everything around us by using our five senses. Information that gets *attended* to would be transferred into our *short-term memory* (STM), the holding place where information in our mind is processed and manipulated. The information may then be stored in our *long-term memory* (LTM) or be used to generate some form of output such as *problem-solving* or *decision-making*.

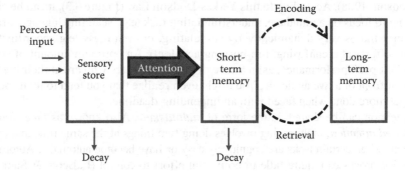

**FIGURE 3.2** Cognition as represented using the information processing model. (Adapted from Wickens, C. D., Hollands, J. G., Banbury, S., & Parasuraman, R., *Engineering psychology & human performance*. East Sussex, UK, Psychology Press, 2015.)

*Attention* is the act of consciously concentrating on information. It is closely related to perception in that attention allows us to focus on what we are perceiving and prevents us from becoming overwhelmed by too much information. For example, you have been paying attention to what is being written on this page and omitting other sounds in the background such as the TV, or cars going by, or people chattering. It will be quite challenging to focus on all these inputs at once. We can thus describe theories of attentional *bottlenecks* (e.g., Broadbent, 1958; Treisman, 1969), which explain how we handle monitoring tasks, be it in air traffic control (Hopkin, 1995), healthcare (e.g., Sanderson, 2006), or process control (e.g., Laberge, Bullermer, Tolsma, & Reising, 2014; Woods, 1995).

Attention can be described in two forms: as visual attention and as a mental resource that fuels cognitive work. *Visual attention* is analogous to a spotlight (Cave & Bichot, 1999; Posner, Snyder, & Davidson, 1980), in which we only consciously process whatever we are consciously seeing (i.e., falls within the light beam). Currently, your visual "spotlight" is on this book, unless you decide to shift it by looking at something else. This concept of *selective attention* (Cowan, 1988; Treisman, 1969) reflects our limited cognitive capacity and duration. We are unable to process everything at the same time and therefore need to be *selective* about what we attend to. This explains why we may be so engrossed with locating the right house while driving that we do not notice the visible pothole on the road. We may also fail to spot the dancing gorilla while counting the times a ball is being passed (Chabris & Simons, 2010), among many other incidents of *inattentional blindness* in which real obvious objects go unnoticed simply because we are busy attending to something else.

Attention can also be described as a *mental resource* that fuels our thinking (Kahneman, 1973). Whenever we pay attention, we consume attentional resources. Driving on a quiet Sunday morning may demand little attentional resource and therefore affords you to engage in a conversation with your friend. Conversely, driving on a busy road in stormy conditions demands a lot more from you, and you tend to feel too overwhelmed to do anything else. Our capacity for attention is limited, and activities that are more demanding thus demand more attentional resources. This capacity decreases when we are bored or overwhelmed (i.e., highly aroused), as described by the Yerkes-Dodson Law (Diamond, Campbell, Park, Halonen, & Zoladz, 2007; Yerkes

& Dodson, 1908). According to this Yerkes-Dodson Law (Figure 3.3), it can be challenging to focus on a mundane, understimulating task (e.g., scanning luggage x-rays) or a task that is very dynamic, overly stimulating, or even noisy (e.g., coordinating rescue efforts at a confusing, fire-burning accident). An optimum amount of stress can elicit peak performance, as in a case of public speakers who are at their best on stage in front of a live audience. You might even realize that you tend to focus better and get more done when faced with an impending deadline.

Attention can be shared, in the form of *multitasking*. Also known as *time-sharing* or *shared attention*, multitasking involves doing two things at the same time and tends to occur when certain tasks are cognitively easy or have become automatic. Automatic cognitive processes require little to no mental effort to control (LaBerge & Samuels, 1974; Schneider & Fisk, 1983). Driving is so automatic to experienced cab drivers that they are able to simultaneously perform other tasks with ease, like engaging in a discussion. Even so, multitasking can still be difficult and this can be explained using the *multiple resources theory* (Wickens, 2002, 2008). This theory states that we have different unique capacities for specific information codes, input modes, and response types. The cab driver is successful at multitasking because driving involves spatial codes, visual input, and manual output, while conversations involve verbal codes, auditory input, and vocal output. Conversely, you may have a hard time simultaneously reading this book and chatting with someone because these activities both use verbal information codes. It is exceptionally hard to do both reading and texting on the phone at the same time, and you likely have to stop one activity in order to focus on the other (Figure 3.4).

There is much to discover about attention, which is a vital component in practically every job and task. A comprehensive breakdown of this topic can be found in *Applied Attention Theory* (Wickens & McCarley, 2008), as well as *The Psychology of Attention* (Pashler & Sutherland, 1998).

STM, occasionally referred to as working memory (Baddeley, 2003; Cowan, 1988), is the mental space in our minds where we perform mental work. We utilize the STM whenever we are consciously thinking, from figuring out the fastest route,

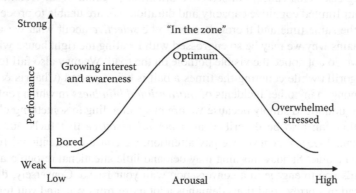

**FIGURE 3.3**  This illustration of the Yerkes-Dodson Law (from Diamond, D. M., Campbell, A. M., Park, C. R., Halonen, J., & Zoladz, P. R., *Neural Plasticity*, 2007, 1–33, 2007; Yerkes, R. M., & Dodson, J. D., *Journal of Comparative Neurology and Psychology*, 18, 459–482, 1908) describes how our arousal affects our task performance.

| | | Information code | Input mode | Output mode |
|---|---|---|---|---|
| | Driving | Spatial | Visual | Manual |
| | Chatting | Verbal | Auditory | Vocal |
| | Reading | Verbal | Visual | Manual |
| | Chatting | Verbal | Auditory | Vocal |
| | Reading | Verbal | Visual | Manual |
| | Texting | Verbal | Visual | Manual |

**FIGURE 3.4** Performing two simultaneous tasks with conflicting information codes, input modes, and/or output modes can result in inefficiencies, according to the Multiple Resources Theory. (From Wickens, C.D., *Human Factors,* 50, 449–445, 2008.)

to performing mental calculations, to recalling the steps to make a peanut butter sandwich. The STM is akin to a work desk, and whatever we pay attention to from the world (perceived) or from the head (recalled) gets temporarily stored on this desk (Figure 3.5). The capacity in this working memory is limited, as identified by Miller's Magical Number 7, Plus or Minus 2: the number of objects an average human can hold in the STM is 7 ± 2 (Miller, 1956). Newer research claims that the STM has a capacity limit of four information *chunks* (Cowan, 2001). Given the precious mental real estate, all information in the STM will decay when not attended

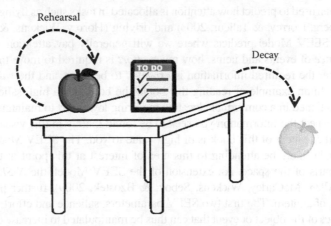

**FIGURE 3.5** The STM is like a work desk that temporarily holds information we are currently attending to.

to, making space for new information (hence its name: "short-term"). About 30% of such information would have been forgotten if left unrehearsed after 3 seconds (Peterson & Peterson, 1959).

The STM is the gateway for information to be stored in the LTM. Information is *encoded* from the STM into the LTM "filing cabinet." Information from the past can also be *retrieved* from the LTM to be actively processed in the STM (Anderson, 1983). The effectiveness of our LTM depends significantly on how well we have encoded information from the STM to the LTM, usually through some form of learning or mental *rehearsal* (Shiffrin & Atkinson, 1967; Atkinson & Shiffrin, 1968). The most effective strategy to encode information into the LTM is *elaborative rehearsal* (Higbee, 1977). During elaborative rehearsal, context and meaning are manipulated into the items in the STM, building connections and drawing relationships. This differs from *maintenance rehearsal*, where information is remembered by constantly repeating it in the mind. When we *recall* information that we have remembered, we transfer information from the LTM back to the STM for mental processing. This process is not always perfect, which is why at times we have that *tip-of-the-tongue* feeling of almost remembering something (Brown, 1991) or recalling something wrong altogether. Our recall ability deteriorates further when we multitask or actively process other information in our STM (Baddeley, Lewis, Eldridge, & Thomson, 1984; Kane & Engle, 2000). For additional insights on memory and its applications, check out *The SAGE Handbook of Applied Memory* (Perfect & Lindsay, 2014).

## METHODS

Various guides and tools exist to help when examining the different components of the information-processing model. Researchers have looked into ways to frame and quantify aspects of attention, STM, and LTM. Such developments aid practitioners when applying cognition in action so as to make sense of the cognitive load involved in task performance, as well as deriving solutions that support cognitive work.

The *SEEV Model of Selective Attention* (Wickens, Helleberg, Goh, Xu, & Horrey, 2001) has been used to predict how attention is allocated in tasks such as flying (Wickens, Goh, Helleberg, Horrey, & Talleur, 2003) and driving (Horrey, Wickens, & Consalus, 2006). The SEEV Model predicts where we will generally pay attention, depending on the *salience* of events and items, how much *effort* is required to move the attention there, whether the required information is *expected* to be there, and the *value* of such information. In an example of reading this book, the book has a high salience and is probably positioned in a comfortable location requiring low effort to maintain attention. You would expect the information you seek to be found mainly in this visual area, and (hopefully) the content of this book is of high value to you. The SEEV Model predicts that you would likely be attending to this area of interest at this point in time rather than other parts of the space. An extension of the SEEV Model, the *N-SEEV Model* (Steelman-Allen, McCarley, Wickens, Sebok, & Bzostek, 2009), further predicts the noticeability of an item. The first two SEEV parameters, salience and effort, are design characteristics of the object or event that can thus be manipulated to increase or decrease the tendency of people selecting it to attend to. The other two parameters, expectancy and value, are driven from past experiences and training, which will influence the likelihood of someone seeking out and paying attention to the item. The tape dispenser, which is well blended into the desk clutter away from your immediate field of view and irrelevant to your current task of reading, will unlikely catch your attention.

When it comes to cognition in action, the STM and its limitation are often associated with cognitive *workload*. A high workload demand would mean there is little or no STM mental capacity to process anything more. As a result, tasks that exceed the available STM capacity would usually not be successful or would have less-than-ideal outcomes. Ways to measure workload include primary task measures, secondary task measures, subjective rating measures, and physiological (or psychophysiological) measures (Meshkati, Hancock, Rahimi, & Dawes, 1995).

*Tracking EEG patterns* is one physiological means of capturing cognitive and working memory load (Berka et al., 2004; Gevins et al., 1998; Klimesch, 1999) and can even identify sections during the cognitive processes that were more demanding (Berka et al., 2007). Another means of assessing STM capacity is by utilizing *secondary* or *subsidiary tasks*, which essentially require the participants to multitask or task switch (Moray, 2013). The secondary task is used to tease out any remaining STM capacity while performing the main or primary task. If the primary task is easy (e.g., riding a bicycle), there should be sufficient STM capacity for a secondary task (e.g., chatting with a friend). Conversely, cognitive overload can be seen when the introduction of the secondary tasks causes performance decrement (i.e., demand for mental resources exceeds available supply). The Multiple-Resource Theory can further anticipate cognitive overload during such dual-tasking situations, particularly when common types of resources are used (Wickens, 2008). The use of subsidiary tasks to assess workload has been performed in driving (Strayer & Johnston, 2001; Zeitlin, 1995) and in simulated process control simulations (Gonzalez, Vanyukov, & Martin, 2005).

## APPLICATION

Practically any task that requires you to use your brain can be mapped to the information-processing model. The model is often illustrated through driving (Lachman, Lachman, Butterfield, 2015; Wickens & Carswell, 2006; Zaidel, Paarlberg, Shinar, 1978) and has also been the basis for studying the negative effects of driving and cell-phone use (Horrey & Wickens, 2006; Strayer, & Drews, 2007; Young, Lee, & Regan, 2008). Specific components of the information-processing model can also be the focus of enhancement to aid task performances. Some examples are presented next.

*Checklists* have consistently been adopted across many high-risk industries as a good strategy to ensure consistency and standardization without relying on LTM. A checklist, such as your shopping list each time you visit your grocer, avoids the need to engage in the vulnerable process of storing and retrieving information to and from your LTM. Checklists are effective at reducing memory-lapse mistakes (Norman, 2013), particularly in aviation (Degani & Wiener, 1990), process control (Plocher, Yin, Laberge, Thompson, & Telner, 2011), and healthcare (Gawande & Lloyd, 2010; Hales & Pronovost, 2006; Walker, Reshamwalla, & Wilson, 2012). In these high-risk industries, a variety of complex procedures and safety protocols exist that can be impossible to remember completely or to recall accurately at the right time. Checklists have helped to ensure that everyone knows the correct protocol to use and the appropriate step to take, so that people do not need to rely on their imperfect memory, especially during time-critical situations. These days, more high-tech devices can be found to serve the same purpose (Herrmann, Brubaker, Yoder, Sheets, & Tio, 1999; Mosier, Palmer, & Degani, 1992).

Many smart phone applications (*apps*) have helped perform complex computation and relieve some of the workload traditionally involved in such tasks. Basic apps like calculators (and regular calculators, of course) are more accurate and reliable than to perform the same task mentally. With Internet search engines (e.g., Google) in every smart phone, knowledge can be accurately and consistently retrieved *from-the-world* instead of *in-the-head*. Navigation software utilizing signals from global positioning satellites (GPS) can automatically derive the best possible routes in foreign lands, some of which may even factor in traffic conditions. These apps have, in turn, been adopted into ridesharing services (e.g., Uber), allowing for almost anyone to pick up paying riders and providing car ride services simply by driving a working vehicle and following instructions from such apps. The cognitive tasks of figuring out where the rider is located, which route to take, and how much to charge have all been automated through the in-built GPS within these apps. Driving apps have become so powerful that taxiing jobs, which were once cognitively intensive, are now made significantly easier (Lee et al., 2015) and have even improved driving performance (Birrell, Fowkes, & Jennings, 2014). With the prevalence of smart phones and the advancement of mobile apps, people will be able to do more with fewer cognitive resources.

## FUTURE TRENDS

The knowledge of cognition is fundamental in many, more advanced topics in human factors. Three topics and applications of cognition relevant in today's working environment are discussed next.

Human cognition and limitations are gaining attention in *error management* and *risk mitigation*. Effective error analysis tools that want to tackle human error would require an understanding of human imperfections. Some examples include the Human Factors Analysis and Classification System (Wiegmann & Shappell, 2003), the Cognitive Reliability and Error Analysis Method (Hollnagel, 1998), and the Technique for the Retrospective and Analysis of Cognitive Errors (TRACer; Shorrock & Kirwan, 2002). When correct root causes are accurately identified, effective solutions could then be designed and developed. This need to mitigate risk by appreciating the human user is particularly evident in today's healthcare (Carayon, 2012; Carayon & Wood, 2010; Gosbee, 2002; Reason, 1995). The U.S. Food and Drug Administration (2000) has even recommended human factors analyses for medical devices prior to market release.

Practically every job requires a need to handle *interruptions* (Bower, Jackson, & Manning, 2015; Jett & George, 2003; Rivera-Rodriguez & Karsh, 2010; Sanderson & Grundgeiger, 2015). Interruptions generally involve brief switching of attention, or even the sharing of attention. Interruptions can disrupt information processing and hinder effective task performance, yet oftentimes, some of these *value-added interruptions* (Rivera, 2014) can be important and time-sensitive. Efforts in *Interruption*

*Science* can be found in the human–computer interaction domain (Bailey & Konstan, 2006; Mark, Gudith, & Klocke, 2008) as well as research work in the field. Many of the recent research efforts have captured the context and effects of interruptions in various domains, and recent work explores interruption management strategies or design (Colligan & Bass, 2012; Kowinsky et al., 2012; Lu et al., 2013).

Another cognitive concept that is frequently employed at work is *prediction* (Griffiths & Tenenbaum, 2006; Orrell, 2007). As another form of cognitive output (e.g., decision making, problem solving), prediction is akin to decision making, but with the time factor projected into the future. Prediction can be top–down or *memory based*, such as anticipating that the postman would routinely arrive at 2 p.m. Prediction can also be bottom–up, or *cue based*, such as reacting to the future location of a tennis ball based on its speed and trajectory. Cue-based predictions can be further enhanced through *predictive displays* (Endsley, Sollenberger, & Stein, 1999; Herdener, Wickens, Clegg, & Smith, 2015; Yin, Wickens, & LaBerge, 2015). While we tend to predict often, the science behind manual prediction is currently very limited still.

## CONCLUSION

The significance of this chapter is simple: as long as you have a brain and use it to process information, you can potentially make mistakes. Having an understanding of how our mind works would help in analyzing root causes in human errors, as well as developing appropriate solutions that do not just tackle our limitations but also complement our cognition in action. The information processing model is a representative and concise framework to map how we think. Within each component of the information processing model lies a vast wealth of knowledge and research that can be explored further beyond the boundaries of this chapter.

## REFERENCES

Anderson, J. R. (1983). Retrieval of information from long-term memory. *Science, 220*(4), 25–30.

Atkinson, R. C., & Shiffrin, R. M. (1968). Human memory: A proposed system and its control processes. *Psychology of Learning and Motivation, 2,* 89–195.

Baddeley, A. (2003). Working memory: Looking back and looking forward. *Nature Reviews Neuroscience, 4*(10), 829–839.

Baddeley, A., Lewis, V., Eldridge, M., & Thomson, N. (1984). Attention and retrieval from long-term memory. *Journal of Experimental Psychology: General, 113*(4), 518.

Bailey, B. P., & Konstan, J. A. (2006). On the need for attention aware systems: Measuring effects of interruption on task performance, error rate, and affective state. *Journal of Computers in Human Behavior, 22*(4). 709–732.

Berka, C., Levendowski, D. J., Cvetinovic, M. M., Petrovic, M. M., Davis, G., Lumicao, M. N., & Olmstead, R. (2004). Real-time analysis of EEG indexes of alertness, cognition, and memory acquired with a wireless EEG headset. *International Journal of Human–Computer Interaction, 17*(2), 151–170.

Berka, C., Levendowski, D. J., Lumicao, M. N., Yau, A., Davis, G., Zivkovic, V. T., & Craven, P. L. (2007). EEG correlates of task engagement and mental workload in vigilance, learning, and memory tasks. *Aviation, Space, and Environmental Medicine, 78*(Supplement 1), B231–B244.

Birrell, S. A., Fowkes, M., & Jennings, P. A. (2014). Effect of using an in-vehicle smart driving aid on real-world driver performance. *IEEE Transactions on Intelligent Transportation Systems, 15*(4), 1801–1810.

Bower, R., Jackson, C., & Manning, J. C. (2015). Quality and safety special issue: Interruptions and medication administration in Critical Care. *Nursing in Critical Care, 20*(4), 183–195.

Broadbent, D. E. (1958). *Perception and communication.* New York: Oxford University Press.

Brown, A. S. (1991). A review of the tip-of-the-tongue experience. *Psychological Bulletin, 109*(2), 20.

Carayon, P. (2012). *Human factors and ergonomics in health care and patient safety* (2nd ed.). Boca Raton, FL: Taylor & Francis.

Carayon, P., & Wood, K. E. (2010). Patient safety: The role of human factors and systems engineering. *Studies in Health Technology and Informatics, 153*, 23–46.

Card, S. K., Moran, T. P., & Newell, A. (1986). The model human processor: An engineering model of human performance. In K. R. Boff, L. Kaufman, & J. P. Thomas (Eds.), *Handbook of perception and human performance* (Vol. 1, pp. 1–35). New York: Wiley.

Cave, K. R., & Bichot, N. P. (1999). Visuospatial attention: Beyond a spotlight model. *Psychonomic Bulletin & Review, 6*(2), 204–223.

Chabris, C., & Simons, D. (2010). *The invisible gorilla: And other ways our intuitions deceive us.* London, UK: HarperCollins.

Colligan, L., & Bass, E. J. (2012). Interruption handling strategies during paediatric medication administration. *BMJ Quality and Safety, 21*(11), 912–917.

Cowan, N. (1988). Evolving conceptions of memory storage, selective attention, and their mutual constraints within the human information processing system. *Psychological Bulletin, 104*(2), 163–191.

Cowan, N. (2001). The magical number 4 in short-term memory: A reconsideration of mental storage capacity. *Behavioral and Brain Sciences, 24*, 87–114.

Diamond, D. M., Campbell, A. M., Park, C. R., Halonen, J., & Zoladz, P. R. (2007). The temporal dynamics model of emotional memory processing: A synthesis on the neurobiological basis of stress-induced amnesia, flashbulb and traumatic memories, and the Yerkes-Dodson law. *Neural Plasticity 2007*, 1–33.

Degani, A., & Wiener, E. L. (1990). *The human factors of flight-deck checklists: The normal checklist (NASA contractor report 177549).* Moffett Field, CA: NASA Ames Research Center.

Endsley, M. R., Sollenberger, R., & Stein, E. (1999). The use of predictive displays for aiding controller situation awareness. *Proceedings of the 43rd Annual Meeting of the Human Factors and Ergonomics Society* (pp. 51–55). Santa Monica, CA: Human Factors and Ergonomics Society.

Gawande, A., & Lloyd, J. B. (2010). *The checklist manifesto: How to get things right.* New York: Metropolitan Books.

Gevins, A., Smith, M. E., Leong, H., McEvoy, L., Whitfield, S., Du, R., & Rush, G. (1998). Monitoring working memory load during computer-based tasks with EEG pattern recognition methods. *Human Factors, 40*(1), 79–91.

Gonzalez, C., Vanyukov, P., & Martin, M. K. (2005). The use of microworlds to study dynamic decision making. *Computers in Human Behavior, 21*(2), 273–286.

Gosbee, J. (2002). Human factors engineering and patient safety. *Quality and Safety in Health Care, 11*, 352–354.

Griffiths, T. L., & Tenenbaum, J. B. (2006). Optimal predictions in everyday cognition. *Psychological Science, 17*(9), 767–773.

Hales, B. M., & Pronovost, P. J. (2006). The checklist—A tool for error management and performance improvement. *Journal of Critical Care, 21*(3), 231–235.

Herdener, N., Wickens, C. D., Clegg, B. A., & Smith, C. A. P. (2015). Mapping spatial uncertainty in prediction. *Proceedings of the 59th Annual Meeting of the Human Factors and Ergonomics Society* (pp. 140–144). Santa Monica, CA: Human Factors and Ergonomics Society.

Herrmann, D., Brubaker, B., Yoder, C., Sheets, V., & Tio, A. (1999). Devices that remind. In F. T. Durso (Ed.), *Handbook of applied cognition* (pp. 377–407). Chichester: John Wiley & Sons.

Higbee, K. L. (1977). *Your memory: How it works and how to improve it*. Englewood Cliffs, NJ: Prentice-Hall.

Hollnagel, E. (1998). *Cognitive Reliability and Error Analysis Method—CREAM*. Oxford: Elsevier Science.

Hopkin, V. D. (1995). *Human factors in air traffic control*. Bristol, PA: Taylor & Francis, Inc.

Horrey, W. J., & Wickens, C. D. (2006). Examining the impact of cell phone conversations on driving using meta-analytic techniques. *Human Factors, 48*(1), 196–205.

Horrey, W. J., Wickens, C. D., & Consalus, K. P. (2006). Modeling drivers' visual attention allocation while interacting with in-vehicle technologies. *Journal of Experimental Psychology: Applied, 12*(2), 67.

Jett, Q. R., & George, J. M. (2003). Work interrupted: A closer look at the role of interruptions in organizational life. *Academy of Management Review, 28*, 494–507.

Kahneman, D. (1973). *Attention and effort*. Englewood Cliffs, NJ: Prentice-Hall.

Kane, M. J., & Engle, R. W. (2000). Working-memory capacity, proactive interference, and divided attention: Limits on long-term memory retrieval. *Journal of Experimental Psychology: Learning, Memory, and Cognition, 26*(2), 336.

Klimesch, W. (1999). EEG alpha and theta oscillations reflect cognitive and memory performance: A review and analysis. *Brain Research Reviews, 29*(2), 169–195.

Kowinsky, A., Shovel, J., McLaughlin, M., Vertacnik, L., Greenhouse, P., Martin, S., & Minnier, T. E. (2012). Separating predictable and unpredictable work to manage interruptions and promote safe and effective work flow. *Journal of Nursing Care Quality, 27*, 109–115.

Laberge, D., & Samuels, S. J. (1974). Towards a theory of automatic information processing in reading. *Cognitive Psychology, 6*, 293–323.

Laberge, J. C., Bullermer, P., Tolsma, M., & Reising, D. V. C. (2014). Addressing alarm flood situations in the process industries through alarm summary display design and alarm response strategy. *International Journal of Industrial Ergonomics, 44*, 395–406.

Lachman, R., Lachman, J. L., & Butterfield, E. C. (2015). *Cognitive psychology and information processing: An introduction*. East Sussex, UK: Psychology Press.

Lee, M. K., Kusbit, D., Metsky, E., & Dabbish, L. (2015). Working with machines: The impact of algorithmic and data-driven management on human workers. In *Proceedings of the 33rd Annual ACM Conference on Human Factors in Computing Systems* (pp. 1603–1612). ACM.

Lu, S. A., Wickens, C. D., Prinet, J. C., Hutchins, S. D., Sarter, N., & Sebok, A. (2013). Supporting interruption management and multimodal interface design: Three meta-analyses of task performance as a function of interrupting task modality. *Human Factors, 55*(4), 697–724.

Mark, G., Gudith, D., & Klocke, U. (2008). The cost of interrupted work: More speed and stress. *Proceedings of the SIGCHI Conference on Human Factors in Computing Systems* (pp. 107–110). ACM.

Meshkati, N., Hancock, P., & Rahimi, M., Dawes, S. M. (1992). Techniques in mental workload assessment. In J. Wilson & E. Corlett (Eds.), *Evaluation of human work. A practical ergonomics methodology* (pp. 605–627). London, UK: Taylor & Francis.

Moray, N. (Ed.). (2013). *Mental workload: Its theory and measurement* (Vol. 8). New York: Springer Science & Business Media.

Miller, G. A. (1956). The magical number seven, plus or minus two: Some limits on our capacity for processing information. *Psychological Review 63*(2): 81–97.

Mosier, K. L., Palmer, E. A., & Degani, A. (1992). Electronic checklists: Implications for decision making. *Proceedings of the Human Factors and Ergonomics Society Annual Meeting* (Vol. 36, No. 1, pp. 7–11). Los Angeles, CA: SAGE Publications.

Norman, D. A. (2013). *The design of everyday things: Revised and expanded edition*. New York, Basic books.

Orrell, D. (2007). *The future of everything: The science of prediction*. New York: Thunder's Mouth Press.

Pashler, H. E., & Sutherland, S. (1998). *The psychology of attention* (Vol. 15). Cambridge, MA: MIT Press.

Perfect, T. J., & Lindsay, D. S. (2014). *The SAGE handbook of applied memory*. London, UK: SAGE Publications.

Peterson, L., & Peterson, M. J. (1959). Short-term retention of individual verbal items. *Journal of Experimental Psychology, 58*(3), 193.

Plocher, T., Yin, S., Laberge, J., Thompson, B., & Telner, J. (2011). Effective shift handover. *Proceedings of the International Conference on Engineering Psychology and Cognitive Ergonomics* (pp. 332–341). Berlin, Heidelberg: Springer.

Posner, M. I., Snyder, C. R. R., & Davidson, B. J. (1980). Attention and the detection of signals. *Journal of Experimental Psychology, 109*, 160–174.

Reason, J. T. (1995). Understanding adverse events: Human factors. In C. A. Vincent (Ed.), *Clinical risk management* (pp. 31–54). London, UK: BMJ.

Rivera, A. J. (2014). A socio-technical systems approach to studying interruptions: Understanding the interrupter's perspective. *Applied Ergonomics, 45*(3), 747–756.

Rivera-Rodriguez, A. J., & Karsh, B. T. (2010). Interruptions and distractions in healthcare: Review and reappraisal. *Quality and Safety in Health Care, 19*, 304–312.

Sanderson, P. (2006). The multimodal world of medical monitoring displays. *Applied Ergonomics, 37*, 501–512.

Sanderson, P. M., & Grundgeiger, T. (2015). How do interruptions affect clinician performance in healthcare? Negotiating fidelity, control, and potential generalizability in the search for answers. *International Journal of Human–Computer Studies, 79*, 85–96.

Schneider, W., & Fisk, A. D. (1983). Attentional theory and mechanisms for skilled performance. In R. A. Magill (Ed.), *Memory and control of action* (pp. 119–143). New York: North-Holland Publishing Company.

Shiffrin, R. M., & Atkinson, R. C. (1967). Storage and retrieval processes in long-term memory. *Psychological Review, 76*, 179–193.

Shorrock, S. T., & Kirwan, B. (2002). Development and application of a human error identification tool for air traffic control. *Applied Ergonomics, 33*(4), 319–336.

Steelman-Allen, K. S., McCarley, J. S., Wickens, C., Sebok, A., & Bzostek, J. (2009). N-SEEV: A computational model of attention and noticing. *Proceedings of the Human Factors and Ergonomics Society Annual Meeting* (Vol. 53, No. 12, pp. 774–778). Thousand Oaks, CA: SAGE Publications.

Strayer, D. L., & Drews, F. A. (2007). Cell-phone-induced driver distraction. *Current Directions in Psychological Science, 16*(3), 128–131.

Strayer, D. L., & Johnston, W. A. (2001). Driven to distraction: Dual-task studies of simulated driving and conversing on a cellular telephone. *Psychological Science, 12*(6), 462–466.

Treisman, A. M. (1969). Strategies and models of selective attention. *Psychological Review, 76(3)*, 282–299.

U.S. Food and Drug Administration (2000). *Medical device use-safety: Incorporating human factors engineering into risk management*. Document Shelf No. 1497. Rockville, MD: Food and Drug Administration.

Walker, I. A., Reshamwalla, S., & Wilson, I. H. (2012). Surgical safety checklists: Do they improve outcomes? *British Journal of Anaesthesia, 109*(1), 47–54.

Wickens, C. D. (2002). Multiple resources and performance prediction. *Theoretical Issues in Ergonomics Science, 3*, 159–177.

Wickens, C. D. (2008). Multiple resources and mental workload. *Human Factors, 50*, 449–455.

Wickens, C. D., & Carswell, C. M. (2006). Information processing. In G. Salvendy (Ed.), *Handbook of human factors and ergonomics* (pp. 111–149). Hoboken, NJ: John Wiley.

Wickens, C. D., Goh, J., Helleberg, J., Horrey, W. J., & Talleur, D. A. (2003). Attentional models of multitask pilot performance using advanced display technology. *Human Factors, 45*(3), 360–380.

Wickens, C. D., Helleberg, J., Goh, J., Xu, X., & Horrey, W. J. (2001). *Pilot task management: Testing an attentional expected value model of visual scanning.* Champaign, IL: Aviation Research Lab, Institute of Aviation.

Wickens, C. D., Hollands, J. G., Banbury, S., & Parasuraman, R. (2015). *Engineering psychology & human performance.* East Sussex, UK: Psychology Press.

Wickens, C. D., & McCarley, J. (2008). *Applied attention theory.* Boca Raton, FL: Taylor & Francis.

Wiegmann, D. A., & Shappell, S. A. (2003). *A human error approach to aviation accident analysis: The Human Factors Analysis and Classification System.* Burlington, VT: Ashgate Publishing, Ltd.

Woods, D. D. (1995). The alarm problem and directed attention in dynamic fault management. *Ergonomics, 38*, 2371–2393.

Yerkes, R. M., & Dodson, J. D. (1908). The relation of strength of stimulus to rapidity of habit-formation. *Journal of Comparative Neurology and Psychology, 18*, 459–482.

Yin, S., Wickens, C. D., Helander, M., & LaBerge, J. C. (2015). Predictive displays for a process-control schematic interface. *Human Factors, 57*(1), 110–124.

Young, K., Lee, J. D., & Regan, M. A. (Eds.). (2008). *Driver distraction: Theory, effects, and mitigation.* Boca Raton, FL: CRC Press.

Zaidel, D. M., Paarlberg, W. T., & Shinar, D., 1978. *Driver performance and individual differences in attention and information processing* (Vol. 1). Driver Inattention. Institute for Research in Public Safety. DOT-HS-803 793. Washington, DC: Department of Transportation.

Zeitlin, L. R. (1995). Estimates of driver mental workload: A long-term field trial of two subsidiary tasks. *Human Factors, 37*(3), 611–621.

## KEY TERMS

**attention:** act of consciously concentrating on information.

**cognition:** cognitive activities involve multiple components (e.g., attention, short-term memory, and long-term memory) interacting with one another as information is being processed in the mind.

**information processing:** receiving information in one form and transforming it into another.

**long-term memory:** "filing cabinet" in our mind where information is *encoded* from the short-term memory and stored for later retrieval.

**multiple resources:** different unique capacities for specific information codes, input modes, and response types.

**short-term memory:** holding place where information in our mind is processed and manipulated; occasionally referred to as working memory.

# 4 Measuring Human Performance in the Field

*Igor Dolgov, Elizabeth K. Kaltenbach,*
*Ahmed S. Khalaf, and Zachary O. Toups*

## CONTENTS

## INTRODUCTION

Human performance is the accomplishment of a task or series of tasks by a person or team of people. Scholars who study human performance seek to understand the requirements, skills, procedures, mechanisms, and outcomes of performance across various domains of skilled human activity, including work, education, warfare, arts, sports, and business. Thus, performance science relies on techniques and methodologies originating in disciplines like psychology, human factors, kinesiology, biology, ergonomics, sociology, economics, marketing, and sports science. Through the evaluation of human performance in a variety of contexts, scientists and engineers can better understand human capabilities in mundane and extreme scenarios, with the goal of enhancing performance through training and technology.

Modeling human performance involves considering a combination of measures and how they are related to one another. Bailey's (1982) model of human performance is a guide for system and interface designers and includes the following parameters: accuracy, efficiency (time), proficiency (skill mastery and acquisition), and (self) satisfaction. It has been applied successfully in many contexts, such as predicting written/typed user input (e.g., Bleha, Slivinsky, & Hussien, 1990; Soukoreff & Mackenzie, 1995) and education (Kearns, 2016). Using this model requires the understanding of the human(s) performing the activity, what the activity is (including what tools, equipage, and crew are needed), and the context in which the activity is performed (e.g., environment, weather, time of day). With these three pieces of information, performance under various operational scenarios can be modeled and understood, with the goal of designing better systems and/or creating a better training regimen for the activities in question.

## FUNDAMENTALS

Human performance depends on the actual task(s) to be performed as well as numerous environmental and human factors (Wickens, Hollands, Banbury, & Parasuraman, 2015). As is expected, the amount of effort needed to complete a task is one of the biggest determinants of performance. Tasks have mental, physical, and personnel demands and both interact to impact human performance. Furthermore, crew composition and individual differences in the performers' age, gender, background knowledge, skills (physical, mental, and communication abilities), affect (mood and emotion), fatigue, and stress can also determine the nature of performance. Additionally, environmental factors, such as background distractions during the task (like noise or other disturbances), lighting (indoors and outdoors), time of day, and weather, can negatively or positively influence workload and, consequently, performance. Thus, measures examining performers' workload, situation awareness, and trust in automated systems go hand-in-hand with directly evaluating human performance.

## METHODS

A number of qualitative and quantitative measures of human performance have been developed.[1] While they employ a large variety of strategies and instruments and measure numerous empirical variables, human performance can be characterized by two main factors: efficacy and efficiency. While *efficacy* (effectiveness) has several operational definitions, typically involving measurements of accuracy and/or errors, it can be simply be understood as a person's *success or failure at performing a given*

---

[1] See Gawron (2008), Meister (2014), and Wilson and Sharples (2015) for reviews.

*task*. Building on efficacy, *efficiency* considers *task completion time* in light of the corresponding successes and/or failures.

Accordingly, human performance methods can be divided into six general categories: accuracy-based measures, time-based measures, task batteries, domain-specific measures, critical incident measures, and team-performance measures (Gawron, 2008). While qualitative measures can be used to glean much information about human performance, the focus of this chapter is on quantitative measures because they lend themselves to statistical analyses. Thereafter, findings from such measures and analyses can be used to inform behavioral models, equipment design, team composition, and training regimen—all in the interest of improving human performance.

## ACCURACY MEASURES

Measures of accuracy can be grouped into two major categories: those that measure correct performance of tasks and those that measure failure or errors. Both major categories contain discrete and scalar metrics. The most typical measures of accuracy that focus on successful accomplishment of an activity or activities include: task(s) completion, percent correct, number correct, correctness score, average score, and probability of correctness (Gawron, 2008). For example, when measuring the impact of dashboard parameters on text readability, Imbeau, Wierwille, Wolf, and Chun (1989) used a percentage correct of read text messages as a measure of accuracy and, unsurprisingly, found that smaller font resulted in a smaller percentage correct of text read by the participants (cf., Gawron, 2008). Since there are generally more ways to fail than to be successful, measures of accuracy that focus on failures are somewhat more numerous and include error rate, error number, percent error, absolute error, relative error, deviations, root-mean-square error, false alarm (detecting a signal when one is *not* present) rate, miss (*not* detecting a signal when one *is* present) rate, and probability of error (Gawron, 2008). For example, when measuring vigilance in a monitoring ("standing watch") task, Galinsky, Warm, Dember, Weiler, and Scerbo (1990) found that false alarms increased significantly as the rate of salient events decreased (cf., Gawron, 2008).

The specific way that accuracy is measured is particular to the task(s) and research question(s) involved. In order to get a better picture of a person's efficacy, many studies examine both successes and failures, and some metrics incorporate both terms in a ratio, such as the ratio of the number correct to the number of incorrect responses. For example, in a typing training task, Ash and Holding (1990) used number of errors divided by the number of correct responses as a measure of accuracy and found significant differences between training methodologies and order effects (cf., Gawron, 2008). Since typing is inherently a process in which mistakes can be corrected and the overall success of the task depends on both the correct typing of letters and the correction of mistakes, utilizing a ratio metric is better than focusing on either mistakes or successes alone.

## TIME MEASURES

Time measures are generally based on duration or speed. Duration measures are used when the research question is about the total time that is takes for a person or team of people to complete a task. Typical measures in this group include time on task, task completion time, response time, duration, looking time, movement time, recognition time, and others (Gawron, 2008). On the other hand, when efficiency is of interest, measures of time can be combined with accuracy metrics to compute speed. Speed and other efficiency metrics are particularly important in the domain of manufacturing, as well as in education and athletics, but less useful for music performance and the arts, in general.

## TASK BATTERIES

Human performance typically involves the balancing of multiple resources to accomplish a number of tasks, often at the same time. Thus, task battery metrics involve the measurement of human performance along two or more tasks performed simultaneously or in series. The motivation behind testing human performance on two or more tasks conducted *in parallel* is that in that real world people rarely perform a single uniform task. Furthermore, the reason for using multiple measures *in series* is that the success completion of complex tasks typically necessitates mastering multiple variables, which may or may not impact each other (Gawron, 2008). Some examples of human performance task batteries are AGARD's Standardized Tests for Research with Environmental Stressors (STRES) Battery (Draycott & Kline, 1996), Armed Forces Qualifications Test (Uhlaner & Bolanovich, 1952), Deutsch and Malmborg Measurement Instrument Matrix (Deutsch & Malmborg, 1982a, 1982b), Work and Fatigue Test Battery (Rosa & Colligan, 1988), and the Unified Tri-Services Cognitive Performance Assessment Battery (UCPAB) (Perez, Masline, Ramsey, & Urban, 1987).

The various components of each task battery are intended to provide a comprehensive understanding of human performance in a given context or scenario. As summarized in Gawron (2008), AGARD's STRES battery is on the smaller end of such metrics and contains seven tests evaluating (1) Reaction time, (2) Mathematical processing, (3) Memory search, (4) Spatial processing, (5) Unstable tracking, (6) Grammatical reasoning, and (7) Dual task (simultaneous) performance of tests 3 and 5. These tests were selected with the criteria of being reliable and valid, as well as quick and easy to administer on widely available computer systems. In contrast, the UCPAB is composed of 25 tests evaluating performance along metrics of spanning the subdomains of memory, decision making, language, auditory and visual processing, and time perception. The collection of metrics selected particularly for this battery included tests that had been validated in previous Department of Defense efforts and showed sensitivity to hostile environments and sustained operations (Gawron, 2008).

## DOMAIN-SPECIFIC MEASURES

Domain-specific measures assess abilities to perform a collection of related tasks that have a common goal or outcome and rely on domain-specific knowledge or skills. For example, in aviation, aircraft parameter metrics include: takeoff and climb measures, cruise measures, approach and landing measures, and various others including composite scores of the aforementioned tests. Other examples come from the domains of industrial and office work, where there are a multitude of measures evaluating performance along engineering, purchasing, management, and planning functions (Gawron, 2008).

## CRITICAL INCIDENT MEASURES

Critical incident measures assess human performance in worst-case scenarios. Any critical incident measure involves the introduction of an event/obstacle or series of events/obstacles that challenges and stresses performers in novel and/or unexpected ways. These measures are particularly useful when safety is paramount and the variables of efficacy and efficiency are secondary. One example summarized in Gawron (2008) is the Critical Incident Technique (Flanagan, 1954), which is a set of specifications for observing the performance of an activity by a person or team. These methodological specifications include the following:

1. Observers (data collectors) must have the following:
   a. Background knowledge of the activity
   b. Some relation to those being observed
   c. Training requirements
2. Groups (or teams) to be observed, including the following:
   a. General description of the group or team
   b. Location
   c. People
   d. Times
   e. Conditions
3. Behaviors to be observed with an emphasis on the following:
   a. General activity type
   b. Specific behaviors
   c. Criteria of relevance to general aim, goal, or outcome
   d. Criteria of importance to general aim, goal, or outcome

## TEAM PERFORMANCE MEASURES

Team (or group) performance measures evaluate the performance of two or more people working in unison on a common goal or set of goals. A key concern in assessing team performance involves considering the impact of communication and coordination on team tasks. The associated costs, in terms of time, cognition, and technological bandwidth, are communication overhead (MacMillan, Entin, & Serfaty, 2004; Toups, Kerne, & Hamilton, 2011). High-performance teams reduce communication overhead through cross-training (Cannon-Bowers et al., 1998; Marks, Sabella, Burke, & Zaccaro, 2002; Volpe, Cannon-Bowers, & Salas, 1996), shared understanding, and body language.

There is a wide variety of team performance measures; most are domain-specific and generally measure group efficacy and/or efficiency. One example noted by Gawron (2008) comes from the military domain and evaluates team performance when accomplishing a command and control (C2) task, which involves the deployment of personnel and resources in a war theater environment. Another example noted by Gawron is Nieva, Fleishman, and Rieck's (1985) Team Dimensions metric, which evaluates team performance when (1) matching resources to task requirements, (2) coordinating responses/communication, (3) pacing the activity, (4) assigning task priority and (5) balancing workload among team members. Another metric, aimed at identifying times of high team performance, is *anticipation ratio*, which assesses the amount of team communication that is devoted to supplying the team with information versus the amount requesting information (Entin & Serfaty, 1999; MacMillan et al., 2004). Information requests contribute to communication overhead, but not situation awareness. Using anticipation ratio requires a work-intensive process of analyzing communication among team members.

## WORKLOAD MEASURES

While often congruent with human performance, workload measures incorporate aspects of performance, perceived difficulty, and expended effort. Workload has previously been defined as effort, activity, accomplishment, and as a set of task demands (Gartner & Murphy, 1979). The task demands are also known as the task load and comprise the total set of goals to be achieved within certain constraints, such as a time or resource limit. The amount of effort expended and the perceived difficulty of the task are influenced by the nature of the specific tasks(s) to be performed as well as the information and equipment provided. Numerous human factors also play a role, including individual differences in performers' background knowledge and experience, adopted strategies, and various affective (emotional) and cognitive styles (Gawron, 2008).

Workload measures fall into four categories: stand-alone (primary task) measures, secondary task measures, subjective measures, and physiological measures. Standalone measures evaluate workload via performance of the specific task that is of interest and measure workload directly, whereas other types of measures are indirect. For example, the Aircrew Workload Assessment System (AWAS) (Reid & Nygren, 1988; cf. Gawron, 2008), directly assesses the performance of an airplane crew at flying its aircraft. On the other hand, secondary task measures,[2] of which there are dozens, assess workload by introducing a secondary (often unrelated) task to the typical operational environment. In this instance, decrements in performance of the secondary task serve as an indirect measure of workload. The third group of measures, of which the National Aeronautics and Space Administration's Task Load Index (NASA-TLX) (Hart, 2006; Hart & Staveland, 1988) is the most widely used, are subjective and rely on performers' ratings of workload using a number of different scales and methods.[3] The fourth group of measures is physiological in nature. Some examples include heart rate, blood pressure, caloric consumption, and electrodermal activity.[4]

As summarized in Table 4.1, each category of workload measures has various determinants and consequent advantages and disadvantages (cf. Gawron, 2008).

---

[2] For a review of secondary-task measures, see Gawron (2008).
[3] For a review of subjective measures, see Gawron (2008).
[4] For a review of physiological measures, see Caldwell (1995).

## TABLE 4.1
## Determinants of Workload Measures

| Measure | Primary Determinant | Secondary Determinant |
|---|---|---|
| Single-task performance | Amount of invested resources | Task difficulty |
| | | Subject's motivation |
| | | Subjective criterion of optimal performance |
| | Resource efficiency | Task difficulty |
| | | Data quality |
| | | Practice |
| Dual-task performance | Amount of invested resources | Task difficulty |
| | | Subject's motivation |
| | | Subjective criteria of optimal performance |
| | Resource efficiency | Task difficulty and/or complexity |
| | | Data quality |
| | | Practice |
| Subjective workload | Amount of invested resources | Task difficulty |
| | | Subject's motivation |
| | | Subjective criteria of optimal performance |
| | Demands on working memory | Amount of time sharing between tasks |
| | | Amount of information held in working memory |
| | | Demand on perceptual and/or central processing resources |

*Source:* Adapted from Gawron, V.J., *Human performance, workload, and situational awareness measures handbook.* Boca Raton, FL: CRC Press, 2008; original source: Yeh, Y., & Wickens, C.D., *Human Factors, 30,* 111–120, 1998.

While stand-alone measures provide a direct measure of workload they are also domain-specific due to their nature and do not generalize very well. Moreover, they may not be sensitive to changes in workload due to unforeseen circumstances. Thus, in order to build a more comprehensive picture of human performance in a variety of operational scenarios, secondary-task, subjective, and/or physiological measures are often used to complement stand-alone measures. However, instruments that do not measure performance in the primary task are not adequate estimates of human workload, since they are greatly impacted by additional variables, such as the demands of the secondary task and individual differences on the parts of the performers.

## SITUATION AWARENESS MEASURES

Successful performance of a given activity depends on possessing the adequate knowledge, skills, and resources to perform that activity. Measures of situation awareness investigate a person's knowledge relevant to the task being performed, including awareness of an ongoing operation or activity, its environmental context, the states of the various human and automated agents involved, and expectations of future state. There are three levels of situation awareness: (1) perception of the elements in the environment, (2) comprehension of the current situation, and (3) projection of future status (Endsley, 1995). Greater knowledge along each of these components generally results in improved performance.

Situation awareness can be assessed directly through measuring knowledge on the fly in real operations as well as in simulations. The most popular direct measure is the Situational Awareness Global Assessment Technique (SAGAT), which involves interrupting performers of a given task and then asking them probing questions (Endsley, 1988). It has been shown to be valid and reliable and can be used in real-world and simulated scenarios (Gawron, 2008). Psychophysiological and subjective measures of situation awareness also exist (e.g., French, Clarke, Pomeroy, Seymour, & Clark, 2007; Koester, 2007; Vidulich, Stratton, Crabtree, & Wilson, 1994). However, as with similar measures of workload, they are indirect and can be influenced by external factors and individual differences on the part of the performers. Situation awareness metrics are particularly useful in the domains of engineering, manufacturing, warfare and peacekeeping, and any operation in which safety is paramount. While also beneficial in athletics, situation awareness metrics have not received much attention in the performance arts.

## Trust Measures

Human performance in the context of using automated systems is also impacted by trust in such systems (Freedy, DeVisser, Weltman, & Coeyman, 2007; Lee & Moray, 1992, 1994; Lee & See, 2004). Lee and Moray (1992) define trust as "the attitude that an agent will help achieve an individual's goals in a situation characterized by uncertainty and vulnerability" (p. 1268). System operators determine their trust in the system by observing the performance in the system and the manner in which the process of accomplishing the goals is understandable. The best performance is produced when the operators know how to appropriately trust and appropriately rely on the automation.

Muir's (1994) model of human trust differentiates between trust, confidence, predictability, and accuracy. Predictability is the factor upon which an operator can make a prediction about future system behavior. The operator has a level of *confidence* in their prediction of what the system will do. The *accuracy* of that prediction can be assessed by comparing the prediction to how the system actually behaved. All of these concepts are distinct, yet related to each other, and are considered when attempting to measure a user's trust in a system.

While trust was traditionally assessed by measuring compliance with automation or with a single question asking operators to rate their trust on some scale, more comprehensive assessments like the Trust in Automated System Scale (TAS; Jian, Bisantz, & Drury, 2000) and Human Computer Trust Scale (HCTS; Madsen & Gregor, 2000) have recently been developed. These surveys utilize a number of items to get a more comprehensive picture of trust. The HCTS contains only positive valence questions, while the TAS contains both negative and positive valence questions. While aggregated scores from both of the measures have been found to be significantly correlated with each other, they are also differentially sensitive to various factors that influence trust (Kaltenbach, 2016).

## APPLICATION

Unmanned aircraft systems (UAS), often called drones, are fixed-wing or rotary-wing aircraft that are piloted remotely by one or more human operators. UAS have primarily been deployed in warfare for the purposes of remote sensing (reconnaissance) and weapons deployment. However, in recent years, UAS have become commercially available and are now being utilized for a variety of contexts, including surveying, disaster response, border protection, search and rescue, agriculture, maintenance, and recreation. Based on the low operating expense and versatility of UAS, it is expected that this industry will experience rapid near-term growth in the civil and commercial sector (North Central Texas Council of Governments, 2011). Recent purchases of UAS manufacturers by tech-giants like Amazon and Google underscore the importance of these technologies in the foreseeable future (Solomon, 2014).

While the word *unmanned* is in their name, UAS require a great amount of human input and control and are prone to accidents. Furthermore, the recent proliferation of this technology is worrisome because UAS experience a much higher rate of mishaps than their manned counterparts do (Dolgov & Hottman, 2011). Thus, efforts are currently being made to quantify and improve human performance when operating UAS. In a civilian context, the focus is currently on safe operations and the integration of UAS into the National Airspace System (Federal Aviation Administration, 2013), whereas in the military realm, the focus is currently on effective and efficient deployment of UAS teams (e.g., U.S. Air Force, 2009).

UAS vary in size; the smallest platforms weigh only a few grams and can fit into one's palm, whereas the biggest platforms are the size of small private jets. Correspondingly, UAS crew composition also varies. In the simplest cases, the entire crew consists of the UAS pilot, who is responsible for the totality of the actions needed to pilot the aircraft safely and effectively. In complex cases, the crew can consist of an external pilot who is responsible for take-off and landing, an internal pilot who is responsible for all other phases of flight, a mission commander who is responsible for mission planning and coordination, a payload operator, and visual observers who are responsible for helping the pilot(s) and mission commander ensure that the UAS is free and clear of other aircraft and parts of the environment. Thus, evaluating human performance in the context of UAS requires a multifaceted approach.

An examination of prior research in the area of unmanned vehicles reveals a number of extant laboratory- and field-based metrics that are already being utilized to measure human performance (e.g., Chen, Hass, & Barnes, 2007), workload (e.g., Moore, Ivie, Gledhill, Mercer, & Goodrich, 2014), situation awareness (e.g., Drury, Riek, & Rackliffe, 2006), trust in automation (e.g., Freedy et al., 2007), and team coordination (e.g., Cooke, Shope, & Kiekel, 2001). In a review of 150 papers examining performance issues that arise when humans operate a robotic vehicle remotely (teleoperation), Chen et al. (2007) identified the following contributing factors that degrade remote perception and robotic-vehicle manipulation: limited field of view, (uncertain) orientation, variable context (camera view point), degraded depth perception, degraded video images, time delay, and motion. These are consistent with the issues also noted in the domain of UAS operations (e.g., Williams, 2004).

Based on the aforementioned factors, Chen et al. (2007) derived a set of recommendations for the design of human–computer interfaces that are used to remotely operate robotic vehicles. These recommendations, such as the use of multimodal displays and controls, focus on creating an immersive interface in which situation awareness and trust are enhanced and workload is minimized. While this is highly desirable, few, if any, current UAS control technologies fit the bill. Starting with the recommendations of Chen et al., researchers and designers can utilize modern wearable-computing technologies to create an immersive UAS control experience and measure human performance and related human factors *during* real-world UAS operations, with the goal of actively adapting to UAS operators' changing needs on-the-fly.

## FUTURE TRENDS

As noted by Roco and Bainbridge (2002), a number of technologies from the domains of biology, medicine, information science, nanoscience, and cognitive science have converged in ways that have enabled meaningful enhancements of human performance. This is particularly evident with the proliferation of automation in our society, which allows for the off-loading of mental and physical workload onto tools and artifacts in the environment. Automation technology is particularly useful in contexts that could potentially endanger human performers, like the use of unmanned aircraft and ground vehicles in warfare. Furthermore, automation has proven useful in contexts that require computational or physical capabilities beyond those of people, including large-scale industrial management, supervisory control, credibility assessment, luggage screening, and clinical/medical decisions (Kaltenbach, Dolgov, & Trafimow, 2016).

Emerging wearable technologies have allowed for the measurement of human performance, workload, situation awareness, and trust in automation on-the-fly during ongoing UAS operations in the field. Moreover, following the design principles outlined by Chen et al. (2007), data gleaned using such technologies can then be fed back into an adaptive UAS control interface. The purpose of adaptive control interfaces is to improve human performance by changing the nature of the human–computer interaction to better fit the users' needs (e.g., Dydek, Annaswamy, & Lavretsky, 2013). With the integration of wearable technologies into UAS control architectures, automation can be used to (a) reduce UAS operators' workload when their task load is high, (b) improve their situation awareness when it is low, (c) aid in planning and decision making, particularly when they are stressed or fatigued, and (d) improve team communication and coordination.

The following technologies, summarized in Table 4.2, include those that have already been utilized in the context of UAS as well as other emerging wearable computing technologies with great potential to be utilized in this domain. Rather than standing alone, these mobile wearable technologies can be utilized in tandem to create an effective adaptive mobile UAS control platform with the potential to greatly enhance human performance and flight safety.

### WEARABLE BIOSENSORS

A number of noninvasive wearable biosensors have recently become commercially available, including those that measure heart rate, blood pressure, blood oxygenation,

**TABLE 4.2**

**Summary of Various Categories of Measures That Can Be Collected with Wearable Technologies**

| Wearable Technologies | Enabled Measures |
| --- | --- |
| Biosensors | Accuracy, time, critical response, workload, situation awareness, trust in automation, critical response |
| Eye-trackers | Accuracy, time, situation awareness, critical response |
| Microphones, headphones, and other acoustic technologies. | Workload, team coordination, critical response, team coordination, critical response |
| Gesture control/measurement interfaces | Workload, critical response |
| GPS and other positioning technologies | Accuracy, time, critical response, team coordination |
| Radio, mobile telephony, and networking | Accuracy, time, critical response workload, situation awareness, trust in automation, team coordination |

respiration rate, body temperature, electrodermal activity (via galvanic skin response), caloric consumption, muscle activity (via electromyography), heart activity (via electrocardiography), and brain activity (via electroencephalography and other brain imaging techniques). Although these technologies are not designed to directly assess human performance, workload, or situation awareness, they are good indicators of expended effort, stress, and fatigue, as well as physiological and mental arousal.

Prior research has shown that these variables are directly related to human performance and workload (e.g., Mehler, Reimer, Coughlin, & Dusek, 2009). So, researchers and designers can utilize various biosensors in concert to measure efficacy, efficiency, workload, and critical incident response. Achieving this goal requires the fusion of data for multiple sensors, but this has been made easier with prior research (Brooks & Iyengar, 1998) and commercially available products like the Zephyr Bioharness, which is a tight-fitting shirt that can be worn underneath UAS operators' regular clothing (Zephyr Bioharness, n.d.).

## MOBILE EYE-TRACKERS

Mobile eye-trackers, like the Tobii Pro Glasses 2 and SMI Eye Tracking Glasses 2, allow for the measurement of people's gaze, eye fixations, and saccades and are usually fitted into a glasses frame or a head-mounted display. Eye tracking techniques have been used to directly and indirectly measure cognition (Gegenfurtner, Lehtinen, & Säljö, 2011) and can easily be worn by UAS crewmembers. This versatile technology can be used to directly assess (sub)task completion, so it can be used for directly measuring efficacy and efficiency, as well as indirectly measuring situation awareness by examining what performers are attending to.

## WEARABLE MICROPHONES, HEADPHONES, AND OTHER ACOUSTIC TECHNOLOGIES

Wearable microphones, headphones, and other acoustic technologies are quite common and essential for the successful operation of UAS. These can be further utilized as measures of arousal, effort, critical response, and team dynamics. Thus, they can provide indices of workload and team coordination. In addition, voice stress analysis can be used to identify whether the speaker is stressed and/or anxious (Meyerhoff, Saviolakis, Koenig, & Yourick, 2001). Moreover, these technologies can be used to assist in flying UAS through voice commands, replacing the need for traditional, stationary control stations.

## WEARABLE GESTURE CONTROL/MEASUREMENT INTERFACES

Wearable gesture control interfaces are also becoming commonplace. In order to measure body and limb position and movement, these technologies rely on worn or held sensor packages that may include some combination of accelerometers, gyroscopes, magnetometers, barometers, and/or inertial measuring units to detect orientation and position as well as electromyography (EMG) biosensors to detect muscle movement. In the context of UAS operations, they can be used to measure workload indirectly and also have the potential to substitute for traditional hand-held controllers that are currently used to pilot most UAS.

## GLOBAL POSITIONING SYSTEM AND OTHER POSITIONING TECHNOLOGIES

All modern UAS platforms utilize a Global Positioning System (GPS) along with a number of other technologies to track the position of the unmanned aircraft while it is in flight. When used in concert with measures like task completion time, these technologies are already being used to evaluate UAS operations efficacy, efficiency, and critical response (Dolgov, 2016). Moreover, in combination with mobile telephony and networking devices, these technologies allow team members to be aware of each other's locations and for scientists to measure team coordination.

## RADIO, MOBILE TELEPHONY, AND NETWORKING

UAS operations already utilize radio, mobile telephony and networking technologies to enable communication among the crew and the aircraft. These technologies can be leveraged for the purposes of integrating the aforementioned wearable technologies for monitoring various aspects of human performance, workload, situation awareness, and trust in automation. In experimental settings, these technologies provide an avenue for the transmission and collection of data. In real-world operation settings, these technologies enable access to computing infrastructures that allow for off-site data processing and decision-making support, which can then be relayed back to a UAS operator with the goals of reducing workload, improving situation awareness, and improving performance. Moreover, since UAS crew composition typically consists of more than one person, these technologies can be used to measure team coherence and coordination. However, due to the inherent limitations of wearable computing devices and communications infrastructure, researchers and designers must be careful to consider potential seams in their various data sharing protocols (Benford et al., 2006; Broll & Benford, 2005; Chalmers & Galani, 2004; Chalmers, MacColl, & Bell, 2003).

## WEARABLE DISPLAYS

Typical UAS control stations rely on laptop or desktop displays to visualize mission data. However, they are bulky and need a significant amount of power, so some mobile UAS control architectures rely on the pilot's vision to determine the location of the aircraft rather than displaying this information on a digital map. With the proliferation of head-mounted displays, it is currently possible to visualize UAS mission data without the need for a traditional control station. Technologies like the Microsoft Hololens (Beloy, 2016) and the Epson Moverio (Epson, 2016) head-mounted displays are already being used in the service of UAS. Many others, like the Vuzix M3000 (Vuzix, 2017), are being used in medical and industrial applications. Furthermore, as flexible display technology progresses (e.g., Zhou et al., 2006), it will be possible to visualize mission-critical data on various parts of the crew's clothing and/or equipage.

## ADAPTIVE CONTROL PLATFORMS: INTEGRATING IT ALL

Traditional UAS control stations have a number of drawbacks, from lack of mobility to imperfectly designed controls. Even when pilots are given optimal resources and training, mishap and accident rates in UAS operations are still relatively high. So, when we consider scenarios in which UAS crews need to be mobile, such as search and rescue, it is evident that traditional control stations are greatly inadequate. These technologies reduce situation awareness because they require the pilot and other members of the crew to interact with artifacts that distract them from what is going on immediately in front of them. Moreover, traditional control stations are entirely insensitive to individual differences and changes in workload, fatigue, and stress, which undoubtedly results in suboptimal human performance when operating these unmanned aircraft.

Alternatively, an adaptive UAS control platform that successfully integrates a wearable display, positioning, networking, communication, and biosensor technologies can positively impact human performance by improving situation awareness, reducing workload, improving trust in automation, and aiding in decision making. The groundwork for this technology has already been done in university research centers like New Mexico State University's PIxL and PACMANe laboratories (e.g., Sharma, Toups, Dolgov, Kerne, & Jain, 2016). Furthermore, Insitu, a drones-as-services company, has recently implemented its Common Open-mission Management Command and Control (ICOMC2) system using a Microsoft Hololens, replacing their laptop-based control station (Beloy, 2016).

# CONCLUSION

Measuring human performance with a high degree of precision is becoming ever more important. While measuring human performance in the field is difficult, emerging wearable-computing technologies are making it possible. These technologies allow scientists to study and measure human behavior and physiology in real time in their actual operating environment. Moreover, they can be used to create adaptive control systems that can maximize human performance by effectively minimizing workload, optimizing situation awareness, and ensuring trust between human and machine components of the system. In the domain of UAS, wearable technologies can be integrated into a mobile, user-centered, adaptive control system that will help pilots and other crewmembers safely and successfully operate these aircraft.

## REFERENCES

Ash, D. W., & Holding, D. H. (1990). Backward versus forward chaining in the acquisition of a keyboard skill. *Human Factors: The Journal of the Human Factors and Ergonomics Society, 32*(2), 139–146.

Bailey, R. W. (1982). *Human performance engineering: A guide for system designers.* Saddle River, NJ: Prentice Hall.

Beloy, J. (2016, April 26). *Insitu to showcase its new commercial business division and latest tech at Xponential 2016.* Retrieved January 13, 2017, from https://insitu.com/press -releases/insitu-to-showcase-its-new-commercial-business

Benford, S., Crabtree, A., Flintham, M., Drozd, A., Anastasi, R., Paxton, M., Tandavanitj, N., Adams, M., & Row-Farr, J. (2006). Can you see me now? *ACM Transactions on Computer-Human Interaction, 13*(1), 100–133.

Bleha, S., Slivinsky, C., & Hussien, B. (1990). Computer-access security systems using keystroke dynamics. *IEEE Transactions on Pattern Analysis and Machine Intelligence, 12*(12), 1217–1222.

Broll, G., & Benford, S. (2005, September). Seamful design for location-based mobile games. In *International Conference on Entertainment Computing* (pp. 155–166). Springer Berlin Heidelberg.

Brooks, R. R., & Iyengar, S. S. (1998). *Multi-sensor fusion: Fundamentals and applications with software.* Saddle River, NJ: Prentice-Hall.

Caldwell, J. (1995). Assessing the impact of stressors on performance: Observations on levels of analyses. *Biological Psychology, 40*(1), 197–208.

Cannon-Bowers, J. A., Salas, E., Blickensderfer, E., & Bowers, C. A. (1998). The impact of cross-training and workload on team functioning: A replication and extension of initial findings. *Human Factors, 40*, 92–101.

Chalmers, M., & Galani, A. (2004). Seamful interweaving: Heterogeneity in the theory and design of interactive systems. In *Proceedings of the 5th conference on Designing interactive systems: Processes, practices, methods, and techniques* (pp. 243–252). New York, NY: Association for Computing Machinery.

Chalmers, M., MacColl, I., & Bell, M. (2003, September). Seamful design: Showing the seams in wearable computing. In *Proceeding of Eurowearable, 2003* (pp. 11–16). Stevenage, UK: Institute of Engineering and Technology.

Chen, J. Y., Haas, E. C., & Barnes, M. J. (2007). Human performance issues and user interface design for teleoperated robots. *IEEE Transactions on Systems, Man, and Cybernetics, Part C (Applications and Reviews), 37*(6), 1231–1245.

Cooke, N. J., Shope, S. M., & Kiekel, P. A. (2001). *Shared-knowledge and team performance: A cognitive engineering approach to measurement* (AFRL-SB-BL-TR-01-0370). Arlington, VA: Air Force Office of Scientific Research.

Deutsch, S. J., & Malmborg, C. J. (1982a). The design of organizational performance measures for human decision making, Part I: Description of the design methodology. *IEEE Transactions on Systems, Man, and Cybernetics, 12*(3), 344–353.

Deutsch, S. J., & Malmborg, C. J. (1982b). The design of organizational performance measures for human decisionmaking, Part II: Implementation example. *IEEE Transactions on Systems, Man, and Cybernetics, 12*(3), 353–360.

Dolgov, I. (2016). Moving towards unmanned aircraft systems integration into the national airspace system: Evaluating visual observers' imminent collision anticipation during day, dusk, and night sUAS operations. *International Journal of Aviation Sciences*, *1*(1), 41–56.

Dolgov, I., & Hottman, S. B. (2011). Human factors in unmanned aircraft systems. In R. K. Barnhart, S. B. Hottman, D. M. Marshall, & E. Shappee (Eds.), *Introduction to unmanned aircraft systems* (pp. 165–180). Boca Raton, FL: CRC Press.

Draycott, S. G., & Kline, P. (1996). Validation of the AGARD STRES battery of performance tests. *Human Factors*, *38*(2), 347–361.

Drury, J. L., Riek, L., & Rackliffe, N. (2006, March). A decomposition of UAV-related situation awareness. In *Proceedings of the 1st ACM SIGCHI/SIGART Conference on Human–Robot Interaction* (pp. 88–94). New York, NY: Association for Computing Machinery.

Dydek, Z. T., Annaswamy, A. M., & Lavretsky, E. (2013). Adaptive control of quadrotor UAVs: A design trade study with flight evaluations. *IEEE Transactions on Control Systems Technology*, *21*(4), 1400–1406.

Endsley, M. R. (1988, May). Situation Awareness Global Assessment Technique (SAGAT). In *Proceedings of the IEEE National Aerospace and Electronics Conference, NAECON 1988* (pp. 789–795). Piscataway, NJ: Institute of Electrical and Electronics Engineers.

Endsley, M. R. (1995). Toward a theory of situation awareness in dynamic systems. *Human Factors*, *37*(1), 32–64.

Entin, E. E., & Serfaty, D. (1999). Adaptive team coordination. *Human Factors*, 41(2), 312–325.

Epson. (2016, September). *Epson partners with DJI to create AR smart glasses solutions for piloting unmanned aerial vehicles.* Retrieved February 21, 2017, from http://global .epson.com/newsroom/2016/news_20160909.html

Federal Aviation Administration. (2013). *Integration of Civil Unmanned Aircraft Systems (UAS) in the National Airspace System (NAS) roadmap.* Retrieved January 13, 2014, from http://www.faa.gov/uas/media/uas_roadmap_2013.pdf

Flanagan, J. C. (1954). The critical incident technique. *Psychological Bulletin*, *51*(4), 327–358.

Freedy, A., DeVisser, E., Weltman, G., & Coeyman, N. (2007, May). Measurement of trust in human–robot collaboration. In *International Symposium on Collaborative Technologies and Systems, 2007. CTS 2007* (pp. 106–114). Piscataway, NJ: Institute of Electrical and Electronics Engineers.

French, H. T., Clarke, E., Pomeroy, D., Seymour, M., & Clark, C. R. (2007). Psychophysiological measures of situation awareness. In J. Noyes, M. Cook, & Y. Masakowski (Eds.). (2007). *Decision making in complex environments* (pp. 291–310). Farnham, UK: Ashgate Publishing.

Galinsky, T. L., Warm, J. S., Dember, W. N., Weiler, E. M., & Scerbo, M. W. (1990). Sensory alternation and vigilance performance: The role of pathway inhibition. *Human Factors: The Journal of the Human Factors and Ergonomics Society*, *32*(6), 717–728.

Gartner, W. B., & Murphy, M. R. (1979). Concepts of workload. In B. O. Hartman & R. E. McKenzie (Eds.), *Survey of methods to assess workload* (No. AGARD-AG-246). Neuilly-sur-seine, France: Advisory Group for Aerospace Research and Development.

Gawron, V. J. (2008). *Human performance, workload, and situational awareness measures handbook.* Boca Raton, FL: CRC Press.

Gegenfurtner, A., Lehtinen, E., & Säljö, R. (2011). Expertise differences in the comprehension of visualizations: A meta-analysis of eye-tracking research in professional domains. *Educational Psychology Review*, *23*(4), 523–552.

Hart, S. G. (2006). NASA-Task Load Index (NASA-TLX); 20 years later. *Proceedings of the Human Factors and Ergonomics Society Annual Meeting*, *50*(9), 904–908.

Hart, S. G., & Staveland, L. E. (1988). Development of NASA-TLX (Task Load Index): Results of empirical and theoretical research. *Advances in Psychology*, *52*, 139–183.

Imbeau, D., Wierwille, W. W., Wolf, L. D., & Chun, G. A. (1989). Effects of instrument panel luminance and chromaticity on reading performance and preference in simulated driving. *Human Factors*, *31*(2), 147–160.

Jian, J. Y., Bisantz, A. M., & Drury, C. G. (2000). Foundations for an empirically determined scale of trust in automated systems. *International Journal of Cognitive Ergonomics*, *4*(1), 53–71.

Kaltenbach, E. K. (2016). *Evaluating the impact of automated aid reliability and transparency on operators' trust: A comparison of the trust in automated systems and human computer trust scales* [unpublished master's thesis]. New Mexico State University.

Kaltenbach, E. K., Dolgov, I., & Trafimow, D. (2016). Automated aids: Decision making through the lens of cognitive ergonomics. In A. C. Sparks (Ed.), *Ergonomics: Challenges applications and new perspectives* (pp. 111–136). New York, NY: Nova Science Publishers.

Kearns, S. K. (2016). *E-learning in aviation*. London: Routledge.

Koester, T. (2007). Psycho-physiological measurements of mental activity, stress reactions and situation awareness in the maritime full mission simulator. In J. Noyes, M. Cook, & Y. Masakowski (Eds.), *Decision making in complex environments* (pp. 311–320). Farnham, UK: Ashgate Publishing.

Lee, J. D., & Moray, N. (1992). Trust, control strategies and allocation of function in human–machine systems. *Ergonomics*, *35*(10), 1243–1270.

Lee, J. D., & Moray, N. (1994). Trust, self-confidence, and adaptation to automation. *International Journal of Human-Computer Studies*, *40*(1), 153–184.

Lee, J. D., & See, K. A. (2004). Trust in automation: Designing for appropriate reliance. *Human Factors*, *46*(1), 50–80.

MacMillan, J., Entin, E. E., & Serfaty, D. (2004). Communication overhead: The hidden cost of team cognition. In E. Salas & S. M. Fiore (Eds.), *Team cognition: Understanding the factors that drive process and performance* (pp. 61–82). Washington, DC: American Psychological Association.

Madsen, M., & Gregor, S. (2000, December). Measuring human-computer trust. In *11th Australasian Conference on Information Systems* (Vol. 53, pp. 6–8), Brisbane, Australia: Australasian Association for Information Systems.

Marks, M. A., Sabella, M. J., Burke, C. S., & Zaccaro, S. J. (2002). The impact of cross-training on team effectiveness. *Journal of Applied Psychology*, *87*(1), 3–13.

Mehler, B., Reimer, B., Coughlin, J., & Dusek, J. (2009). Impact of incremental increases in cognitive workload on physiological arousal and performance in young adult drivers. *Transportation Research Record: Journal of the Transportation Research Board*, *2138*, 6–12.

Meister, D. (2014). *Human factors testing and evaluation* (Vol. 5). Amsterdam, the Netherlands: Elsevier Science Publishers.

Meyerhoff, J. L., Saviolakis, G. A., Koening, M. L., & Yourick, D. L. (2001). Physiological and biochemical measures of stress compared to voice stress analysis using the Computer Voice Stress Analyzer (CVSA) (No. DODPI95-P-0032). Fort Jackson, SC: Department of Defense Polygraph Institute.

Moore, J., Ivie, R., Gledhill, T., Mercer, E., & Goodrich, M. (2014, March). Modeling human workload in unmanned aerial systems. In *AAAI Spring Symposium Series: Formal Verification and Modeling in Human-Machine Systems* (pp. 44–49). Palo Alto, CA: Association for the Advancement of Artificial Intelligence.

Muir, B. M. (1994). Trust in automation: Part I. Theoretical issues in the study of trust and human intervention in automated systems. *Ergonomics*, *37*(11), 1905–1922.

Nieva, V. F., Fleishman, E. A., & Rieck, A. (1985). *Team dimensions: Their identity, their measurement and their relationships*. Bethesda, MD: Advanced Research Resources Organization.

North Central Texas Council of Governments. (2011). *Unmanned aircraft systems report*. Retrieved June 18, 2012, from http://www.nctcog.org/aa/jobs/trans/aviation/plan/Unmanned AircraftSystemsReport.pdf

Perez, W. A., Masline, P. J., Ramsey, E. G., & Urban, K. E. (1987). *Unified tri-services cognitive performance assessment battery: Review and methodology*. Dayton, OH: Systems Research Labs, Inc.

Reid, G. B., & Nygren, T. E. (1988). The subjective workload assessment technique: A scaling procedure for measuring mental workload. *Advances in Psychology*, *52*, 185–218.

Roco, M. C., & Bainbridge, W. S. (2002). Converging technologies for improving human performance: Integrating from the nanoscale. *Journal of Nanoparticle Research*, *4*(4), 281–295.

Rosa, R. R., & Colligan, M. J. (1988). Long workdays versus rest days: Assessing fatigue and alertness with a portable performance battery. *Human Factors: The Journal of the Human Factors and Ergonomics Society*, *30*(3), 305–317.

Sharma, H. N., Toups, Z. O., Dolgov, I., Kerne, A., & Jain, A. (2016). Evaluating Display Modalities Using a Mixed Reality Game. In *Proceedings of the 2016 Annual Symposium on Computer-Human Interaction in Play* (pp. 65–77). New York, NY: Association for Computing Machinery.

Solomon, B. (2014, March 4). Facebook follows Amazon, Google into drones with $60 million purchase. *Forbes*. Retrieved May 1, 2014, from http://www.forbes.com/sites/briansolomon /2014/03/04/facebook-follows-amazongoogle-into-drones-with-60-million-purchase/

Soukoreff, W. R., & Mackenzie, S. I. (1995). Theoretical upper and lower bounds on typing speed using a stylus and a soft keyboard. *Behaviour & Information Technology*, *14*(6), 370–379.

Toups, Z. O., Kerne, A., & Hamilton, W. A. (2011). The Team Coordination Game: Zero-fidelity simulation abstracted from fire emergency response practice. *ACM Transactions on Computer–Human Interaction*, *18*(23), 1–37.

Uhlaner, J. E., & Bolanovich, D. J. (1952). *Development of Armed Forces qualification test and predecessor army screening tests, 1946–1950* (No. PRS-976). Washington, DC: U.S. Army, Adjutant General's Office.

U.S. Air Force. (2009). *Unmanned aircraft systems flight plan 2009–2047*. Retrieved August 16, 2013, from http://www.fas.org/irp/program/collect/uas_2009.pdf

Vidulich, M. A., Stratton, M., Crabtree, M., & Wilson, G. (1994). Performance-based and physiological measures of situational awareness. *Aviation, Space, and Environmental Medicine*, *65*(5), A7–A12.

Volpe, C. E., Cannon-Bowers, J. A., & Salas, E. (1996). The impact of cross-training on team functioning: An empirical investigation. *Human Factors*, *38*, 87–100.

Vuzix. (2017, January). Vuzix unveils award-winning M3000 smart glasses at CES 2017. Retrieved January 13, 2017, from http://ir.vuzix.com/press-releases/detail/1535/vuzix -unveils-award-winning-m3000-smart-glasses-at-ces-2017

Wickens, C. D., Hollands, J. G., Banbury, S., & Parasuraman, R. (2015). *Engineering psychology & human performance*. Hove, UK: Psychology Press.

Williams, K. W. (2004). *A summary of unmanned aircraft accident/incident data: Human factors implications* (No. DOT/FAA/AM-04/24). Oklahoma City, OK: Federal Aviation Administration, Civil Aeromedical Institute.

Wilson, J. R., & Sharples, S. (Eds.). (2015). *Evaluation of human work*. Boca Raton, FL: CRC Press.

Zephyr BioHarness. (n.d.). Retrieved January 13, 2017, from http://www.zephyr-technology .nl/en/product/71/zephyr-bioharness.html

Zhou, L., Wanga, A., Wu, S. C., Sun, J., Park, S., & Jackson, T. N. (2006). All-organic active matrix flexible display. *Applied Physics Letters*, *88*(8), 083502.

## KEY TERMS

**adaptive control:** the capability of a system to modify its operation to achieve the best possible performance given a dynamically changing environment and/or demands on the system operator.

**human performance:** the accomplishment of a task or series of tasks by a person or team of people.

**trust (in automation):** the attitude that an automated system component will function in accordance with the operator's expectations and behavior.

**unmanned aircraft system (UAS):** sometimes called a drone, a UAS is a remotely piloted aircraft and all the associated supporting technologies.

**wearable computers:** also known as body-borne computers or wearables; consist of a variety of mobile electronic devices that are worn under or on top of clothing.

**workload:** physical and/or mental effort and activity.

# 5 Situation Awareness in Sociotechnical Systems

## Nathan Lau and Ronald Boring

## CONTENTS

## INTRODUCTION

Situation awareness (SA), simply defined as "knowing what's going on" (Endsley, 1995b), has been a major subject of research since Endsley (1988a) formally brought forth the notion into the human factors community. Over the past three decades, SA conjures up a level of attention in both research and practice (Patrick & Morgan, 2010; Rousseau, Tremblay, & Breton, 2004) that firmly establishes the importance and relevance of the notion. It is difficult to imagine a case when workers of socio-technical systems can consistently function effectively without much awareness of the situation. After all, we step onto a plane expecting that the air traffic control-ler has a "good picture" of the air space to land and take off aircrafts (Jeannot, 2000); undergo a surgery expecting that the anesthesiologist is tracking the vital signs to maintain homeostasis (Gaba, Howard, & Small, 1995; McIlvaine, 2007); take a taxi expecting that the driver understands the traffic environment and signs to maneuver the vehicle (Gugerty, 1997); and send soldiers to a war expecting that the commander comprehends the status of the artillery and the intelligence about the enemies to direct the troops (Bryant, Lichacz, Hollands, & Baranski, 2004; Riley, Endsley, Bolstad, & Cuevas, 2006). The literature contains numerous studies of SA in many other application domains, such as sports (James & Patrick, 2004), nuclear power (Lau, Jamieson, & Skraaning, 2012b), infantry (French, Matthews, & Redden, 2004), and cyber security (Brynielsson, Franke, & Varga, 2016). There is no short-age of evidence to indicate that the SA notion resonates with both researchers and practitioners across many domains.

Much SA research in human factors and engineering centers on the develop-ment of a new human performance dimension for measurements.[1] SA is not a simple construct—SA is not a basic emotion that can be measured physiologically; it is not a personality trait that holds constant. SA is dynamic, and it is a complex phenomenon that emerges from multiple aspects of perception and cognition. There is no simple, unambiguous symptom that a human has SA or not. Yet, SA permeates everything a person does and affects the quality of their decision making and actions. Researchers and practitioners measure SA from different perspectives—to indicate whether workers possess the competence to search for critical information and to understand the operating conditions or whether the user interfaces (or the systems, in general) are adequately designed to support information search and comprehension by work-ers. For example, the U.S. Nuclear Regulatory Commission (NRC) requires utilities seeking an operating license of a nuclear power plant to measure SA for validating human factors engineering of the main control room (O'Hara, Higgins, Fledger, & Pieringer, 2012). Simply put, SA matters for safety and usability. SA research is important because it refines psychological measurement in the quest to find the best insights into human performance.

The significance of SA in measurement is well established given the number of publications on SA research. However, the perspectives in SA theory and mea-surement are diverse in the literature (cf., three special journal issues on SA—the *Journal of Human Factors* in 1995, *Theoretical Issues of Ergonomics Science* in 2010, and *Cognitive Engineering and Decision Making* in 2015). To make sense of the diverse SA research, this chapter presents a pragmatic treatment of SA theories and measurements based on our experience of measuring SA in the nuclear domain. Our intention is to highlight some key considerations and our insights for practitio-ners and starting researchers attempting to measure SA in their application domains. The remainder of this chapter begins with fundamentals of SA, reviewing several prominent SA theories. The chapter then turns to reviewing the common methods for measuring SA. After the review, we present our experience in the application of SA for measuring human performance and evaluating control technology in the nuclear industry and conclude with future trends in SA research and applications. Although the featured domain is nuclear power, the conclusions on SA easily gener-alize to other safety critical industries, from aerospace to oil and gas, military, and manufacturing.

## FUNDAMENTALS

The literature contains many theoretical approaches to SA, and some rooted in strong psychological research paradigms remain highly influential (see Endsley,

---

[1] Endsley (2011) has applied the SA to create design guidance, but the majority of the research attention on SA focuses on measurements.

2004; Fracker, 1991; Jeannot, 2000; Rousseau et al., 2004; Salmon et al., 2008). These SA theories or models have substantial influence on the formulation of many SA measures and are tremendously useful for informing the use of SA measurements. This section reviews the major SA theories with a focus on their distinctive contributions, omitting the criticisms and subtle distinctions mentioned in the literature. This position is adopted for two reasons. First, available space for this chapter precludes full description of individual theories, let alone all the supporting and opposing arguments brought forth by different researchers. Readers interested in the details should refer to the extensive references for original arguments and detailed critiques. Second, and more importantly, the focus on the distinct characteristics between theories can highlight their complementary strengths that would likely be of highest interest to practitioners applying SA for measurements in their domains. Details on unresolved scientific issues in SA are important and interesting but often do not directly address real-world problems. In focusing on the respective strengths of individual theories and measures, this chapter aims to guide practitioners in deciding what SA theories and measures to adopt for their applications (see Stanton, Salmon, & Walker, 2015; Stanton, Salmon, Walker, Salas, & Hancock, 2017).

## INFORMATION PROCESSING APPROACH TO SA

Endsley (1995b) adopts the *information processing model* to define SA as perception, comprehension, and projection of elements in the environment. The information processing model treats cognition like a computer—there are inputs, processing, and outputs—yet with human components—environmental inputs, sensory systems, perception, short-term memory cognition, long-term memory, response selection, and response execution. Endsley conceives SA as three types of knowledge products that shape response selection and execution within the information processing model. Endsley's concept explicitly leaves all the cognitive processes (i.e., situation assessment) generating SA to be addressed by the information processing model. The most important distinction of this work is the derivation of information processing based on SA definition.[2]

## TASK ANALYSIS APPROACH TO SA

Patrick et al. (Patrick & James, 2004; Patrick & Morgan, 2010) adopted hierarchical task analysis (HTA) to model SA. A common human factors engineering technique, HTA (Annett, 2004; Stanton, 2006) decomposes complex tasks into subtasks to identify what system information and controls should be made available to the operators. Patrick et al. presented a simple HTA to decompose Endsley's three levels of SA into goals and subtasks for attaining perception, comprehension, and projection. The HTA approach thus converges toward the Goal-Directed Task Analysis utilized

---

[2] Endsley's three-level conceptualization does provide a more precise definition than "knowing what's going on." However, perception, comprehension, and projection of elements in the environment are not formal or exact definitions. For example, perception is a subdiscipline of psychology. Most other definitions of SA are also not formal.

to inform Endsley's Situation Awareness Global Assessment Technique (SAGAT; Endsley, 1995a). Patrick et al. present a method useful for examining SA-related tasks, without adopting a particular definition or model of SA. The most important distinction of this work is using task analysis to model the tasks relevant to SA.

## PERCEPTUAL-ACTION CYCLE APPROACH TO SA

Adams, Tenney, and Pew (1995) and Smith and Hancock (1995) adopted the perceptual-action cycle to model SA. The perceptual-action cycle (Neisser, 1976) suggests that people possess generic representations or knowledge templates for different types of situations, known as schemas (Bartlett, 1932), and gain awareness by filling in the templates with the situational details. Further, people must engage in a cycle of acting on the environment to gather situational information and using that situational information to guide their actions. In essence, SA is viewed as the results of periodically updating a generic knowledge template with situational information. The most important distinction of this work is the explicit integration of actions and environment into acquiring SA, emphasizing the impact of processes themselves on SA.

## COGNITIVE WORK ANALYSIS/ECOLOGICAL APPROACH TO SA

Flach et al. (Flach, Mulder, & van Paassen, 2004; Flach & Rasmussen, 1999) adopted the Abstraction Hierarchy (Rasmussen, 1985) and Decision Ladder (Rasmussen, 1986) in Cognitive Work Analysis (Vicente, 1999) to model SA. The Abstraction Hierarchy represents the situation with structural means–ends relationships, which have been shown to be psychologically relevant for problem solving. The Decision Ladder represents awareness of the worker based on an adapted version of the information processing model that explicitly includes cognitive shortcuts. Flach et al. thus present the separate tools for modeling situation and awareness. The most important distinction of this work is the explicit prescription to model the domain or "ecology" for describing the situation.

## DISTRIBUTED COGNITION APPROACH TO SA

Stanton et al. (Salmon, Stanton, Walker, & Jenkins, 2009; Stanton et al., 2006) adopted the *propositional network* to develop the distributed SA (DSA) concept. In DSA, the propositional network represents the knowledge within and interactions between different human and machine components across the system. The proposition network is able to describe how SA at the system or organization level can emerge from different components exchanging knowledge about the situation with one another. The most important distinction of DSA is the prescription of modeling SA distribution across a complex system as a network, highlighting the "transactive" nature of SA in a system.

These five SA theories capture the most prominent theoretical approaches. As reviewed, these SA theories have important distinctions and emphases on what aspects of the operator or system to model. Thus, practitioners and researchers may adopt a theory most suitable for their application needs rather than seeking for absolute truth or best all-around approach. Table 5.1 matches the general application interests to the individual SA theories, highlighting how different SA theories can serve different purposes. For example, practitioners seeking a system evaluation should select DSA for generating a network assessment depicting the information acquisition, transformation and transmission capabilities, and deficiencies across human and machine components. The emphasis of DSA is not on individual workers. To focus on individuals, Endsley's three levels of SA might be suitable for air traffic control when workers need to interpret many pieces of information in the time-space dimension under time pressure. In essence, the various SA theories provide different utilities in practice.

Understanding the different theoretical approaches to SA can direct practitioners and researchers toward the most important aspects of the systems and operator work. Most importantly, adopting a particular theory helps guide the development or selection of SA measures. Though commonly lacking exact instruction for developing measurement methods, SA theories postulate the content and structure of user knowledge about the situation (e.g., Endsley's three levels of SA), thereby providing guidance on how to customize measurement techniques in human factors for measuring SA.

## TABLE 5.1
## Application Merits of Various SA Models

| SA Theory | Defining Paradigm | Application Merit |
|---|---|---|
| Endsley (1995b, 2000) | Information processing model | When applications need to focus on interpreting a significant amount of relatively unambiguous information similar to computers (i.e., the application domain has a well-defined problem space such as a spatial-temporal one in air traffic control) |
| Patrick and James (2004) | Task analysis | When applications need to focus on explicitly identifying operator behaviors on searching and processing information |
| Adams et al. (1995), Smith and Hancock (1995) | Perception-action cycle and schema theory | When applications need to focus on eliciting knowledge, memory structure, and behavioral strategies of experts |
| Flach et al. (2004), Flach and Rasmussen (1999) | Ecological approach and cognitive work analysis | When applications need to focus on modeling complex systems for supporting/understanding problem solving or interpreting ambiguous information |
| Stanton et al. (2006) | Distributed cognition and network analysis | When applications need to focus on transmission and transformation of information between system components (e.g., military command and control in which many actors play a role in transformation and transmission of information) |

## METHODS

The greatest significance of SA research is the implications on human performance measurement. Efforts in theoretical research contribute to unique customization of common measurement techniques in human factors to assess operator SA across a wide range of studies and applications. Endsley (1988b, 1995a, 1995b) was first to take this general approach by formulating the three levels of SA and applying her theory into SAGAT. Endsley's SA model and measure are still widely adopted in research and practice. Nevertheless, since the introduction of SAGAT, many researchers have applied SA characterizations to customize human factors techniques into specific SA measures. SA measures generally fall into one or a combination of five generic human factors measurement techniques:

- *Probe-based*: Ask users questions (about the situation) either in real time or during a pause/break.
- *Subjective rating scale*: Administer a rating scale (on items regarding their knowledge about the situation) either in real time or during a pause/break.
- *Expert/observer rating*: Ask an expert to observe users and rate performance (in terms of situation assessment/awareness).
- *Process/physiological based*: Collect behavioral data of the user activities (that indicate acquisition of knowledge about the situation, such as eye gaze on relevant areas of the display).
- *Performance based*: Infer user knowledge about the situation based on their control actions (e.g., redirecting aircrafts, shutting a valve) and system parameters (e.g., parameters within safety range, no accidents).

The literature contains several discussions on the advantages and disadvantages of these generic human factors measurement techniques for measuring SA (see Charlton & O'Brien, 2002; Endsley, 1995a; Pew, 2000). Given their relative merits, all these techniques have been adopted to develop SA measures. Table 5.2 presents a brief assessment and some prominent SA measures for individual techniques.

As with theory, the measurement discussion on SA includes a number of alternatives that appear complementary to one another at the generic technique level even though different researchers may advocate their own method(s) much more strongly than others. For example, the probe-based technique is the most advocated for SA measurements. However, some researchers opt for expert ratings because administering probes can be too intrusive to examine SA in highly realistic work settings (Jeannot, Kelly, & Thompson, 2003). The relative merits of the various measurement techniques and different requirements of the application domains consequently lead to many different SA measures. In addition to the few domain-general measures listed in Table 5.2, the literature also contains multiple SA measures for individual domains (Fracker, 1991; Jeannot et al., 2003; Salmon, Stanton, Walker, & Green, 2006).

In essence, selecting SA measures to evaluate technology or operators should be based on not only reliability and validity but also the suitability of the measurement

**TABLE 5.2**

**Generic Human Factors Measurement Techniques and Examples of SA Measures**

| Generic Technique | Examples of SA Measures | General Advantage | General Disadvantage | Main Customization Efforts to Measure SA |
|---|---|---|---|---|
| Probe based | SAGAT (Endsley, 1995a) SPAM (Durso & Dattel, 2004) | –Relate intuitively to the SA notion (i.e., declarative knowledge of the situation) | –Require probes (also known as queries) that may be intrusive to work activities and limited in what can be practically elicited<br>–Limited in generalization as probes are often context dependent and predefined (i.e., less dynamic) | –Design probes, usually system if not scenario specific, to produce reliable and valid SA scores<br>–Design minimally intrusive probe administration methods |
| Self-rating | SART (Taylor, 1990) | –Use same questionnaires for every study, minimizing cost and improving generalization | –Confound with confidence or other self-assessment dimensions (i.e., lack of reference in reality)<br>–Fail to cater uniquely to specific situational characteristics (in scenario-based testing) | –Design SA-specific questionnaire items and anchors for classic psychometric testing of reliability and validity |

(Continued)

## TABLE 5.2 (CONTINUED)
### Generic Human Factors Measurement Techniques and Examples of SA Measures

| Generic Technique | Examples of SA Measures | General Advantage | General Disadvantage | Main Customization Efforts to Measure SA |
|---|---|---|---|---|
| Expert rating | None for general domains but many for specific domains such as SA/BARS for aviation (Matthews & Beal, 2002) | –Use same rating scale for every study, minimizing cost<br>–Permit experts to account for the unique actions taken by individual workers supporting dynamic representation of situation | –Rely on availability of experts and their competence in observing indication of human performance (and potentially their hindsight of the simulation scenarios) | –Design SA-specific questionnaire items and anchors for classic psychometric testing of reliability and validity<br>–Train experts in rating methods |
| Process based | Area of Interest (AOI) with eye-gaze (e.g., VISA by Drøivoldsmo et al., 1998) | –Produce continuous measurements | –Require substantial (subjective) interpretation of the process indices to relate to knowledge about the situation (e.g., "look but do not see" phenomenon) | –Define (or prove) the relationship between process indices and specific dimensions of SA |
| Performance based | Production level, control action errors | –Cost minimally to implement (if available)<br>–Provide criterion validity highlighting direct implications on productivity and safety | –Confound with other system performance dimensions or require substantial (subjective) interpretation of performance indices to relate to knowledge about the situation | –Define (or prove) the relationship between performance indices and specific dimensions of SA |

technique for the application domain. In many cases, the suitability of the techniques can have major impacts on measurement outcome. The next section presents a case study of developing and using SA measures specific to nuclear process control, emphasizing the importance of domain properties in applications. The case study aims to highlight the usefulness of applying SA in a pragmatic manner.

## APPLICATION

Concerns over emissions from electricity generation and the volatility of fossil fuel prices have led to recent and steady investments into the nuclear industry. In the United States, many nuclear power plants are undergoing modernization to extend their operating life cycles. These modernization efforts result in control rooms mixed with analog and digital technology (i.e., "hybrid control room") from different decades. Such significant changes in existing control rooms can greatly impact operators who have trained and worked on the old technology. Around the world, particularly in China as well as the United States, new nuclear power plants are being constructed to add or replace electric power capacity. These new plants are built on the latest process control and computer technology, leading to new digital control room designs that may significantly shift the nature of operator work familiar to the utilities and regulators. Finally, small-modular reactors (SMRs) are being designed to provide scalable and easy-to-manufacture nuclear power plants. SMR plants are expected to have a completely new concept of operations that prescribe a few operators controlling many reactors in collaboration with automation (as opposed to a few operators controlling a large-scale reactor). Such new concepts of operations in the nuclear industry clearly require iterative design and evaluation cycles to arrive at the optimal human–automation collaboration. All these new designs and upgrades for nuclear power plant control rooms require integrated system validation (ISV)—high-fidelity simulator testing with professional operators as mandated by the regulators—to ensure adequate human factors engineering and public safety. In the United States, the NRC specifies SA as a critical human performance dimension that must be supported by the control room and thus demonstrated in the ISV (O'Hara et al., 2012).

To support the nuclear industry on modernization and new construction, we have developed SA measures specific to nuclear process control and used those measures in several full-scope nuclear power plant simulator studies. The purpose of these simulator studies is to assess the expected benefits and potential short falls of various user interfaces and technology for the control room. The assessment results inform practitioners and regulators on what technology to adopt for their modernization projects and new builds.

To select and develop the SA measures for assessing control room technology, we first reexamined process plant operations and available control room field studies to understand how operators develop SA, given the complexity of nuclear power plants. This process helped us to make sense of the diverse SA research for developing measures specific to the nuclear domain (Lau, Jamieson, & Skraaning, 2013). As a result of this research effort, we have characterized SA according to the situation assessment activities of monitoring, reasoning/diagnosis, and self-regulation engaged by nuclear power plant operators (Lau et al., 2012b; Lau & Skraaning, 2015). These domain-specific characterizations of SA provided a basis for customizing generic measurement techniques (Table 5.2) for our simulator studies.

We have customized three generic techniques to measure SA based on our examination into how nuclear power plant operators assess situations. To evaluate the knowledge derived from monitoring process plants, we customized the probe/query-based technique to develop the Process Overview Measure (Lau, Jamieson, & Skraaning, 2016a; Lau, Jamieson, & Skraaning, 2016b). We also customized an expert-rating technique to develop Halden Open Probe Elicitation (HOPE; Burns et al., 2008; Skraaning et al., 2007) and Automation Scenario Understanding Rating Scale (ASURS; Lau, Jamieson, & Skraaning, 2012a) to assess operators in their quality of reasoning and diagnosis. In other words, HOPE and ASURS help quantify expert judgment on how well operators understand the operating circumstances. Finally, we combined the expert- and self-rating techniques to develop a metacognitive accuracy measure to assess self-regulation (Lau, Skraaning, & Jamieson, 2009). We have adopted three different techniques because our research on SA in process control indicates that the characteristics of each subdimension are uniquely suited to specific measurement techniques.

For the purpose of illustrating the application of SA, here, we further describe the Process Overview Measure and its findings from two full-scope simulator studies. We applied the probe-based technique for developing the Process Overview Measure, which evolved from SAGAT and the Situation Awareness Control Inventory (Hogg, Follesø, Volden, & Torralba, 1995) for the process control domain. The first step of the Process Overview Measure was talking with process experts to identify the key process parameters at specific time points in the simulator test scenarios. We then used these process parameters for creating the queries. Queries served as questions about specific parameters that were important during certain plant evolutions. For example, it is important to know the reactor power level when synchronizing the turbine generator to the electrical grid because the optimal power level ensures that the generator load can match grid demands. Querying operators to their awareness of this specific parameter ensured that responses were a meaningful and relevant reflection of the situation that operator must manage. In addition to administering the queries to operators at the prespecified time points of the scenarios, we asked the simulator operator or process expert supporting the studies to answer the same queries for comparison with the answers of the operators.

In two simulator studies, the Process Overview Measure provided valuable assessment of control room technology. The first study investigated the merits of Ecological Interface Design (EID; Vicente & Rasmussen, 1992) in representative nuclear process control settings. In this study, the Process Overview Measure indicated that the licensed operators were more aware of key process parameters with ecological displays than conventional single-sensor-single-indicator displays and state-of-practice mimic displays (Burns et al., 2008). Further, the Process Overview findings corroborated with the task performance results, adding confidence to the benefits of EID for nuclear process control.

The second study investigated the merits of overview displays for controlling advanced automation by licensed operators with varying levels of familiarity with the nuclear power plant design. In this study, advanced automation was programmed to behave similarly to an autopilot in commercial aircrafts. This kind of advanced automation design will become common in the nuclear industry, particularly for SMRs. The Process Overview Measure revealed that operators recruited from a nuclear power plant with similar processes to the simulator's had better awareness of key parameters than operators recruited from plants with different processes. This finding suggested that the Process Overview Measure can be useful to assess operator qualification. The Process Overview Measure also revealed that operators had better awareness of key parameters with the overview displays, but task performance results indicated that operators were less effective in executing control actions with the overview displays (Skraaning, Eitrheim, & Lau, 2010). Debriefing revealed that the operators might have been overly focused on monitoring the automated process rather than actively engaging in corrective actions. This finding suggested that further research on user interface design was necessary for managing advanced automation efficiently. Specifically, the overview displays were effective at supporting SA but not task performance. The results of the study also illustrated the importance of including multiple measures to assess human performance holistically.

## FUTURE TRENDS

Over the last several decades, automation is becoming more complex and pervasive for improving productivity and safety across many domains. As automation assumes the responsibility for routine tasks and well-known events, operators are left primarily in a monitoring role that may decrease SA until faced with tackling rare or unanticipated events. An attentional shift is required at this point, while managing hard-to-understand technology under stressful circumstances (Bainbridge, 1983). The trend of increasing automation will likely continue; hence, future SA research will naturally need to model and measure awareness of automation for diagnosing design and training inadequacies. In light of increasing automation, future SA research will also need to include self-awareness (Lau et al., 2009), which has been shown to affect the use of automation. SA research may also serve the crucial design role of developing automation displays that help operators maintain SA even when their role is primarily passive or process monitoring. Maintaining SA is important for times when automation fails, so that decisions can be made promptly when the operator suddenly and really needs SA.

Recently, wearable biosensors (e.g., eye-trackers, heart-rate monitors, brain activity sensors) have improved drastically in both functionality and cost. Wearable biosensors can provide continuous and real-time physiological measurements, whereas most other types of SA measurements can provide only periodic data points. However, physiological measurements require substantial inference to indicate whether operators possess good SA. For example, an eye-tracker can only indicate whether someone has looked at a relevant system parameter but cannot determine whether that person actually has registered the parameter value into memory or awareness for decision making (i.e., "the look but does not see" phenomenon). So, future research will need to investigate what and how data from various wearable biosensors can be "fused" together to infer operator SA reliably.

## CONCLUSION

SA is a complex notion that has motivated development of many theories and measures. The perspectives on SA are diverse in the literature, but careful examination of individual theories and measures highlights their complementary nature in existing research. Thus, we recommend that a pragmatic approach to selecting SA theories and measurements is most productive for applications. Specifically, we believe that practitioners should find the theory and measurement methods most suitable to worker activities and system operating characteristics in their domains. From our research effort to inform the nuclear industry on control room design, we illustrate that SA measures are useful in assessing operator qualification and user interface effectiveness. In essence, SA is useful in the applications of assessing technology and workers. This chapter illustrates that the merits of SA applications can be realized with careful selection of SA theories and measures based on the nature of the application domains. SA is rich not only in theory but also in application. This chapter aims to help see the practical application anchored in SA theory.

## ACKNOWLEDGMENTS

Nathan Lau is indebted to Dr. Jamieson of the University of Toronto, Canada, and to Dr. Skraaning Jr. of the Institute for Energy Technology, Norway, for his graduate research training. The authors are also grateful for the feedback on earlier versions of the manuscript from the editors as well as the graduate students of the Virginia Cognitive Engineering Laboratory.

## REFERENCES

Adams, M. J., Tenney, Y. J., & Pew, R. W. (1995). Situation awareness and the cognitive management of complex systems. *Human Factors, 37*(1), 85–94.

Annett, J. (2004). Hierarchical task analysis (HTA). In N. A. Stanton, A. Hedge, K. Brookhuis, E. Salas, & H. Hendrick (Eds.), *Handbook of human factors and ergonomics methods* (pp. 33-31–33-37). Boca Raton, FL: CRC Press.

Bainbridge, L. (1983). Ironies of automation. *Automatica, 19*(6), 775–779.

Bartlett, F. C. (1932). *Remembering: A study in experimental and social psychology.* Cambridge, UK: Cambridge University Press.

Bryant, D. J., Lichacz, F., Hollands, J. G., & Baranski, J. V. (2004). Modeling situation awareness in an organizational context: Military command and control. In S. P. Banbury & S. Tremblay (Eds.), *A cognitive approach to situation awareness: Theory and application* (pp. 104–116). Hampshire, UK: Ashgate.

Brynielsson, J., Franke, U., & Varga, S. (2016). Cyber situational awareness testing. In B. Akhgar & B. Brewster (Eds.), *Combatting cybercrime and cyberterrorism: Challenges, trends and priorities* (pp. 209–233). Cham, Switzerland: Springer International Publishing.

Burns, C. M., Skraaning Jr., G., Jamieson, G. A., Lau, N., Kwok, J., Welch, R., & Andresen, G. (2008). Evaluation of ecological interface design for nuclear process control: Situation awareness effects. *Human Factors, 50*, 663–679.

Charlton, S. G., & O'Brien, T. G. (Eds.). (2002). *Handbook of human factors testing and evaluation.* Mahwah, NJ: Lawrence Erlbaum Associates.

Drøivoldsmo, A., Skraaning Jr., G., Sverrbo, M., Dalen, J., Grimstad, T., & Andresen, G. (1998). *Continuous measures of situation awareness and workload* (HWR-539). Retrieved from Halden, Norway.

Durso, F. T., & Dattel, A. R. (2004). SPAM: The real-time assessment of SA. In S. P. Banbury & S. Tremblay (Eds.), *A cognitive approach to situation awareness: Theory and application* (pp. 137–154). Hampshire, UK: Ashgate.

Endsley, M. R. (1988a). Design and evaluation for situation awareness enhancement. *Proceedings of the 32nd Annual Meeting of the Human Factors and Ergonomics Society*, 97–101.

Endsley, M. R. (1988b). *Situation Awareness Global Assessment Technique (SAGAT)*. Paper presented at the National Aerospace and Electronics Conference (NAECON), New York, NY, USA.

Endsley, M. R. (1995a). Measurement of situation awareness in dynamic systems. *Human Factors, 37*(1), 65–84.

Endsley, M. R. (1995b). Toward a theory of situation awareness in dynamic systems. *Human Factors, 37*(1), 32–64.

Endsley, M. R. (2004). Situation awareness: Progress and directions. In S. P. Banbury & S. Tremblay (Eds.), *A cognitive approach to situation awareness: Theory and application* (pp. 342–351). Hampshire, UK: Ashgate.

Flach, J. M., Mulder, M., & van Paassen, M. M. (2004). The concept of the situation in psychology. In S. P. Banbury & S. Tremblay (Eds.), *A cognitive approach to situation awareness: Theory and application* (pp. 42–60). Hampshire, UK: Ashgate.

Flach, J. M., & Rasmussen, J. (1999). Cognitive engineering: Designing for situation awareness. In N. B. Sarter & R. Amalberti (Eds.), *Cognitive engineering in the aviation domain* (pp. 153–179). Mahwah, NJ: Erlbaum.

Fracker, M. L. (1991). *Measures of situation awareness review and future directions* (AL-TR-1991-0128). Retrieved from Dayton, OH: http://www.dtic.mil/cgi-bin/GetTRDoc?Location=U2&doc=GetTRDoc.pdf&AD=ADA262672

French, H. T., Matthews, M. D., & Redden, E. S. (2004). Infantry situation awareness. In S. P. Banbury & S. Tremblay (Eds.), *A cognitive approach to situation awareness: Theory and application* (pp. 254–274). Hampshire, UK: Ashgate.

Gaba, D. M., Howard, S. K., & Small, S. D. (1995). Situation awareness in anesthesiology. *Human Factors: The Journal of the Human Factors and Ergonomics Society, 37*(1), 20–31.

Gugerty, L. J. (1997). Situation awareness during driving: Explicit and implicit knowledge in dynamic spatial memory. *Journal of Experimental Psychology: Applied, 3*(1), 42–66. doi:10.1037/1076-898x.3.1.42

Hogg, D. N., Follesø, K., Volden, F. S., & Torralba, B. (1995). Development of a situation awareness measure to evaluate advanced alarm systems in nuclear power plant control rooms. *Ergonomics, 38*(11), 2394–2413.

James, N., & Patrick, J. (2004). The role of situation awareness in sports. In S. P. Banbury & S. Tremblay (Eds.), *A cognitive approach to situation awareness: Theory and application* (pp. 297–316). Hampshire, UK: Ashgate.

Jeannot, E. (2000). *Situation awareness, synthesis of literature research* (ECC Note No. 16/00). Retrieved from Brussels, Belgium: https://www.eurocontrol.int/eec/gallery/content/public/document/eec/report/2000/031_Situation_Awareness_Literature_Search.pdf

Jeannot, E., Kelly, C., & Thompson, D. (2003). *The development of situation awareness measures in ATM systems* (HRS/HSP-005-REP-01). Retrieved from Brussels, Belgium: https://www.eurocontrol.int/sites/default/files/content/documents/nm/safety/safety-the-development-of-situation-awareness-measures-in-atm-systems-2003.pdf

Lau, N., Jamieson, G. A., & Skraaning Jr., G. (2012a). *Inter-rater reliability of expert-based performance measures.* Paper presented at the Proceedings of the 8th American Nuclear Society International Topical Meeting on Nuclear Plant Instrumentation & Control and Human–Machine Interface Technologies (NPIC & HMIT), San Diego, CA.

Lau, N., Jamieson, G. A., & Skraaning Jr., G. (2012b). Situation awareness in process control: A fresh look. *Proceedings of the 8th American Nuclear Society International Topical Meeting on Nuclear Plant Instrumentation & Control and Human-Machine Interface Technologies (NPIC & HMIT)*, 1511–1523.

Lau, N., Jamieson, G. A., & Skraaning Jr., G. (2013). Distinguishing three accounts of Situation Awareness based on their domains of origin. *Proc. of the 52nd Annual Meeting of the Human Factors and Ergonomics Society*, 220–224.

Lau, N., Jamieson, G. A., & Skraaning, G. (2016a). Situation awareness acquired from monitoring process plants—The Process Overview concept and measure. *Ergonomics, 59*(7), 976–988. doi:10.1080/00140139.2015.1100329

Lau, N., Jamieson, G. A., & Skraaning, G. (2016b). Empirical evaluation of the Process Overview Measure for assessing situation awareness in process plants. *Ergonomics, 59*(3), 393–408. doi:10.1080/00140139.2015.1080310

Lau, N., & Skraaning Jr., G. (2015). *Exploring sub-dimensions of situation awareness to support integrated system validation.* Paper presented at the Proceedings of the 9th American Nuclear Society International Topical Meeting on Nuclear Plant Instrumentation & Control and Human-Machine Interface Technologies (NPIC & HMIT), Charlotte, NC.

Lau, N., Skraaning Jr., G., & Jamieson, G. A. (2009). *Metacognition in nuclear process control.* Paper presented at the Proc. of the 17th Triennial World Congress on Ergonomics, Beijing, China.

Matthews, M. D., & Beal, S. A. (2002). *Assessing situation awareness in field training exercises* (Research Report 1795). Retrieved from U.S. Army Research Institute for the Behavioral and Social Sciences: http://www.dtic.mil/cgi-bin/GetTRDoc?Location=U2&doc=GetTRDoc.pdf&AD=ADA408560

McIlvaine, W. B. (2007). Situational awareness in the operating room: A primer for the anesthesiologist. *Seminars in Anesthesia, Perioperative Medicine and Pain, 26*(3), 167–172. doi:http://dx.doi.org/10.1053/j.sane.2007.06.001

Neisser, U. (1976). *Cognition and reality: Principles and implications of cognitive psychology.* San Francisco: Freeman.

O'Hara, J. M., Higgins, J. C., Fledger, S. A., & Pieringer, P. A. (2012). *Human factors engineering program review model* (NUREG-0711 Rev. 3). Retrieved from Washington, DC: https://www.nrc.gov/docs/ML12324A013.pdf

Patrick, J., & James, N. (2004). A task-oriented perspective of situation awareness. In S. P. Banbury & S. Tremblay (Eds.), *A Cognitive approach to situation awareness: Theory and application* (pp. 61–81). Hampshire, UK: Ashgate.

Patrick, J., & Morgan, P. L. (2010). Approaches to understanding, analysing and developing situation awareness. *Theoretical Issues in Ergonomics Science, 11*(1), 41–57.

Pew, R. W. (2000). The state of situation awareness measurement: Heading toward the next century. In M. R. Endsley & D. J. Garland (Eds.), *Situation awareness analysis and measurement* (pp. 33–47). Mahwah, NJ: Lawrence Erlbaum Associates.

Rasmussen, J. (1985). The role of hierarchical knowledge representation in decision making and system management. *IEEE Trans. Systems, Man and Cybernetics, SMC-15*, 234–243.

Rasmussen, J. (1986). *Information processing and human-machine interaction: An approach to cognitive engineering.* New York: Elsevier Science Publishing Co. Inc.

Riley, J. M., Endsley, M. R., Bolstad, C. A., & Cuevas, H. M. (2006). Collaborative planning and situation awareness in Army command and control. *Ergonomics, 49*(12–13), 1139–1153. doi:10.1080/00140130600612614

Rousseau, R., Tremblay, S., & Breton, R. (2004). Defining and modeling situation awareness: A critical review. In S. P. Banbury & S. Tremblay (Eds.), *A cognitive approach to situation awareness: Theory and application* (pp. 3–21). Hampshire, UK: Ashgate.

Salmon, P. M., Stanton, N. A., Walker, G. H., Baber, C., Jenkins, D. P., McMaster, R., & Young, M. S. (2008). What really is going on? Review of situation awareness models for individuals and teams. *Theoretical Issues in Ergonomics Science, 9*(4), 297–323.

Salmon, P. M., Stanton, N. A., Walker, G. H., & Green, D. (2006). Situation awareness measurement: A review of applicability for C4i environments. *Applied Ergonomics, 37*(2), 225–338.

Salmon, P. M., Stanton, N. A., Walker, G. H., & Jenkins, D. P. (2009). *Distributed situation awareness: Theory, measurement and application to teamwork*. Hampshire, UK: Ashgate.

Skraaning Jr., G., Eitrheim, M. H. R., & Lau, N. (2010). *Coping with automation in future plants*. Paper presented at the 7th American Nuclear Society International Topical Meeting on Nuclear Plant Instrumentation, Control and Human-Machine Interface Technologies (NPIC&HMIT), Las Vegas, NV.

Skraaning Jr., G., Lau, N., Welch, R., Nihlwing, C., Andresen, G., Brevig, L. H., Veland, Ø., Jamieson, G. A., Burns C., & Kwok, J. (2007). *The ecological interface design experiment (2005)* (HWR-833). Retrieved from Halden, Norway.

Smith, K., & Hancock, P. A. (1995). Situation awareness is adaptive, externally directed consciousness. *Human Factors, 37*(1), 137–148.

Stanton, N. A. (2006). Hierarchical task analysis: Developments, applications, and extensions. *Applied Ergonomics, 37*(1), 55–79. doi:http://dx.doi.org/10.1016/j.apergo.2005.06.003

Stanton, N. A., Salmon, P. M., & Walker, G. H. (2015). Let the reader decide: A paradigm shift for situation awareness in sociotechnical systems. *Journal of Cognitive Engineering and Decision Making, 9*(1), 44–50. doi:10.1177/1555343414552297

Stanton, N. A., Salmon, P. M., Walker, G. H., Salas, E., & Hancock, P. (2017). State-of-science: Situation awareness in individuals, teams and systems. *Ergonomics, 60* (4), 449–466.

Stanton, N. A., Stewart, R., Harris, D., Houghton, R. J., Baber, C., McMaster, R., Salmon, P. et al. (2006). Distributed situation awareness in dynamic systems: Theoretical development and application of an ergonomics methodology. *Ergonomics, 49*(12–13), 1288–1311. doi:10.1080/00140130600612762

Taylor, R. M. (1990). Situation awareness rating technique (SART): The development of a tool for aircrew systems design. In: *AGARD Conference Proceedings No. 478, Situational awareness in aerospace operations* (pp. 3/1–3/17). Neuilly Sur Seine, France: NATO-AGARD.

Vicente, K. J. (1999). *Cognitive work analysis: Toward safe, productive, and healthy computer-based work*. Mahwah, NJ: Lawrence Erlbaum Associates.

Vicente, K. J., & Rasmussen, J. (1992). Ecological interface design: Theoretical foundations. *IEEE Transactions on Systems, Man and Cybernetics, 22*(4), 589–606.

## KEY TERMS

**cognitive work analysis:** an analysis framework for modeling sociotechnical systems in five phases: work domain analysis, control task analysis, strategies analysis, social organization and cooperation analysis, and work competencies analysis.

**distributed cognition:** an approach that conceptualizes human cognition as distributed across objects, tools, and individuals in the environment rather than confined to an individual.

**ecological interface design:** an interface design framework that emphasizes analysis of the work domain or environment for specifying design requirements in addition to analysis of the user as emphasized in user-centered design.

**ecological psychology:** a branch of psychology that emphasizes the richness of information in the environment for perception and action, as opposed to computations in the human brain.

**information processing model:** a framework that conceptualizes human cognition as several processing stages: sensation, perception, short-term memory, long-term memory, attention, decision making, and control actions.

**perceptual action cycle:** a framework that conceptualizes human cognition or understanding to emerge from continuous cycles of perceiving and acting on the environment.

**probe-based measures:** performance measures that administer short questions known as probes to collect data from users while they are performing some tasks or during a pause for simulated tasks.

**process overview measure:** a probe-based measure developed specifically for process control to measure situation awareness, specifically on the awareness of parameter changes in process plants.

**situation awareness:** a human performance concept for describing what users or workers need to know for effective decision making.

# Section II

## System and Environmental Considerations

# 6 Automation in Sociotechnical Systems

*Stephen Rice and Rian Mehta*

## CONTENTS

## INTRODUCTION

Technology has become an accepted partner to our daily lives (MacKay & Vogt, 2012). Automation, in turn, has been integrated into almost every task, process, and piece of equipment we interact with. With the aid of automation, tasks have become easier and, in some cases, have eliminated human workload altogether (Endsley, 1999). This is not without its limitations and pitfalls, but the benefits of automation in today's world far outweigh these issues 10-fold (Bolton & Bass, 2010). Just settling on a definition of automation can be tricky as well, as it could be a variation of a number of things based on the context. One fairly universal definition states that automation is "the accomplishment of work or a mechanical or electrical task that otherwise would need to be accomplished by a human operator" (Wickens & Hollands, 2000, p. 377). To fully understand where we are going, it is imperative that we understand the foundation that has helped us to this stage.

Research in the field of automation has been extensive over the past few decades (Onnasch, Wickens, Li, & Manzey, 2013). Early research focused on outlining the concepts and introduced a plethora of new terminology in order to do so. As the field has grown, so has the complexity increased in order to find more intricate explanations to more detailed questions. One of the more salient topics that come up in discussions regarding automation is that of automation dependency and the nature in which an operator's performance is impacted by their trust and reliance in the automation (Endsley, 1996). The question is still largely debated by experts: "Does increased automation reduce operator situational awareness, and thereby impact safety?"

In this chapter, we will look at a couple of different aspects of the broad and expansive term *automation*. We will primarily focus on the fundamentals of automation and the building blocks of automation research that have led us to today, following which we will delve into exploring some of the unique and fascinating applications of the same. To wrap up, we will summarize some future trends we are witnessing, from industries around the world, which could quite literally change our lives.

## FUNDAMENTALS

Gaining a comprehension of the entire field of automation is a mammoth task that does not seem feasible for most practitioners. In order to provide a cursory overview of the most noticeable and significant works for the automation literature, certain key ideas and works are discussed in this section. These fundamentals represent the basic and broad topic area of automation and allow for the practitioner to have a baseline of reference when dealing with the automation literature.

### CLASSIFICATION OF AUTOMATION

To begin, one of the earliest scientific papers on automation dealt with outlining this new field, defining key terminology, and providing a reference for future expansion of the concept. Sheridan and Verplank (1978) proposed a classification of the levels of automation where the upper levels suggested the bulk of the control lay with the automation, whereas in the lower tiers, the control and decision making was primarily left to the human with some circumstances involving the automation's assistance. The 10 levels were as follows:

- Level 1: The human operator does the task with no assistance and no input from the automation.
- Level 2: The computer aids in the decision process by putting forth a set of viable options.
- Level 3: The computer evaluates all options and suggests the most viable. The human operator must then choose to follow the recommendation or not.
- Level 4: The computer only presents one option and one alternative. The human operator decides which one should be done.
- Level 5: The computer executes the most viable option only after the human operator approves the action.
- Level 6: The computer selects the action and informs the human operator who must veto the decision within a certain amount of time; otherwise, the action is implemented by the computer.
- Level 7: The computer does the action and tells the human operator what was done.
- Level 8: The computer does the action and tells the human only if the human operator asks.
- Level 9: The computer does the action but only tells the human operator if it is deemed necessary to inform them of the same.

- Level 10: The computer does the action, acts autonomously, and ignores the human completely.

Continuing with the concept of categorizing and differentiating automation, Parasuraman, Sheridan, and Wickens (2000) expanded with outlining four stages of automation. The model works based on the concept of integrating automation into the manner in which people process information. In this model of automation, the four stages of human–machine interactions are information acquisition, information analysis, decision selection, and action implementation. A new area of automation that does not strictly conform to these stated levels or stages of automation is that of adaptive automation. In traditional automation setups, the characteristics of the system/operation are predetermined along with the nature of the automation use. Adaptive automation, however, accounts for changes in the stage, level, or type of automation used as the operation is in progress (Scerbo, 1996). It involves a dynamic sharing and allocation of tasks between the human and the automation in order to increase safety and efficiency (Byrne & Parasuraman, 1996). The system at this point can help determine when workload on the operator is high and accordingly assign more tasks to the automation in order to alleviate or balance the load or, conversely, assign more tasks to the human during periods of extremely low workload, thereby reengaging the operator in the system.

Some issues to address prior to adaptive automation implementation include determining what aspect of the task will be included in said adaptation, what manner in which the need for adaptation will be determined or measured, and, lastly, who would be responsible for making the decision to adapt. In terms of what to adapt, the most common practice would be to select the choice that has the greatest change in workload. Factors used to determine when to engage the adaptive automation include estimates of environmental influences, measurements of operator performance, and measurements of workload. With respect to the last issue of who should decide, this still remains open with arguments for and against each side. Are the human operators capable of correctly and unambiguously assessing the level of workload and need for adaptation during operations?

## THE HUMAN–AUTOMATION RELATIONSHIP

Parasuraman and Riley (1997) talked about the relationship that humans share with automation in terms of use, misuse, disuse, and abuse. In this context, automation use refers to the standard and appropriate amount of engaging and disengaging automation to aid in reducing workload or increasing efficiency. However, misuse refers to a state of overreliance or dependency on the automation that can have hazardous implications. Overreliance in the automation can occur due to a state of complacency. This state of complacency goes hand in hand with the presence of excessive levels of trust in the automation. As trust increases in the automation, so does the state of reliance in the automation (Lee & See, 2004). The most prevalent hazardous implication of this situation is a lack of supervision over the automation's actions, which in automation failure scenarios could lead to a level of unpreparedness that can cause catastrophic results.

On the opposite spectrum of overreliance and excessive trust is the concept of disuse that occurs due to a lack of trust in the automation to perform accurately or efficiently. The operator chooses to avoid the use of the automation as trust has been negatively influenced by past experiences of automation inaccuracy, primarily in the form of false alerts. The reasons for automation implementation usually center around increased performance, economic benefits, or increased safety. However, in certain cases, automation may be designed or implemented without much consideration of the operator, and this can lead to inefficient, unnecessary automation—the implementation of automation for the sake of implementing automation. This concept is called automation abuse.

Building off of the concepts of overtrust, research conducted by Meyer (2001) introduced a model of reliance and compliance as they relate to operator responses in conditions with different warnings. The results of the study showed an interesting aspect of the relationship between types of warnings and the type of relationship with the automation (warning system). When there were hazard warnings, participants tended to exhibit compliance characteristics; that is, the system notifies the operator of an issue, thereby indicating a need for corrective action, and the operator has to decide to comply with said warning by checking the status of the system in question. Conversely, an "all clear" message elicits more of a reliance behavior from the operator; that is, the alert states that the system is in a safe state, and the operator can choose to rely on the automation that the system is in fact in a safe state. In general, what has been postulated is that operator compliance is more influenced by false alarms of the automation, whereas operator reliance is influenced by the automation missing certain cues. The study of warnings and alerts has been a major topic associated with the research in the automation field.

A study by Rice (2009) delved further into the theories involved in analyzing the operator's trust in automation. Participants were asked to conduct a task to identify certain signals and were assisted by an automated aid of varying reliability. Unsurprisingly, as the reliability of the aid increased, the performance of the human–automation paring increased. Additionally, the study also showed that an increase in false alarms results in a decrease in compliance, while an increase in automation misses resulted in a decrease in reliance. The study concluded that in target present/target absent trials, the type of error differentially affects performance, implying that automation bias (false alarms or misses) has differential effects on performance.

Merritt and Ilgen (2008) added to the body of literature regarding trust in automation with the expansion that individual differences in user personality and perceptions of the machine have an influence on this relationship. The authors postulate that perceptions and schemas created based on propensities to trust had significant influences on trust in the machine. For instance, individuals with higher propensities to trust that were paired with initially reliable machines have a stronger schema for reliable performance and therefore expect the automation to perform correctly. Therefore, when the automation does not perform correctly, these failures are more likely to be noticed and remembered. The study shows that it is important to consider individual differences when establishing an overarching model for human–automation interactions. This study showed that individual operator differences accounted for 52% of variance in trust above the effects of the characteristics of the machine.

When dealing with automation and the advanced capabilities of some technologies today, the disadvantages that have cropped up need addressing. The issues with complacency and overreliance have been touched on in this section; however, another aspect to this complex phenomenon is that of situational awareness. Research has shown that as reliability of the automation increases, participants are likely to succumb to latent errors due to increased complacency (Bailey & Scerbo, 2007). The discussions of automation in the realms of the aviation industry state that the purpose of the same is to decrease workload, thereby allowing pilots to think ahead and analyze situations better, but oftentimes hazardous situations are not avoided due to the inefficient use of this extra time afforded by the automation (Chow, Yortsos, & Meshkati, 2014). This brings up an issue of lack of situational awareness due to overreliance in or overdependence on automation. In the aviation industry, and others, a decrease in manual skill sets may also be observed due to repeated use or dependence on automation. This combination of decreased skills, decreased situational awareness, and complacency is termed as *OOTLUF syndrome* or *out of the loop unfamiliarity* (Wickens, 1992). The argument that arises due to the potential consequences of OOTLUF is whether the automation implementation benefits outweigh the potential hazards that it creates. Mentioned earlier was the concept of adaptive automation, which could prove to be a concept whose application could alleviate the loss of situational awareness as the system would adjust the number of tasks assigned to the human and the automation based on the current workload. The safety concerns being discussed predominantly focus on aviation autopilot applications, but as we will see in the next sections, autopilot styled features are beginning to emerge in other sectors as well, primarily automobiles. These issues could have similar implications on the future of automobile operations as well, and could lead to certain potentially hazardous situations.

## METHODS

With a firm understanding of the fundamentals of human factors, the next step is the implementation of this newfound knowledge. In order to do this, we will outline some of the methods and tools that human factors professionals and researchers can potentially utilize to further their work. At our disposal is a wide gamut of interesting and extremely useful technological devices and tools. Some of the most cutting-edge technologies popular in the human factors field are virtual and augmented reality simulators, eye tracking devices, and several types of virtual reality wearable headsets.

Aviation research has been using static and full motion simulators for many years, with an entire line of research in that field. However, new and innovative research is beginning to emerge utilizing virtual and augmented reality simulators. Simulating emergency scenarios for various types of jobs is becoming an extremely effective method of personnel training, not only in aviation but also in several different fields, one of which is law enforcement and personal security. Billinghurst, Clark, and Lee (2015) summarize the last 50 years of augmented reality research in their survey of the topic area.

Another useful tool that human factors professionals are using more and more is eye tracking tools such as the one made by Tobii. Researchers are always intrigued with eye tracking data, as these devices can perform some amazing functions, such as eye positions, movement, movement patterns, length of gaze, and many more. As is evident, these functions can be extremely helpful in conducting applied research that can help many different fields to understand human attention and behavior. Eye tracking research has been around for over a decade, with several researchers interested in the monitoring performance and attention span of pilots in this current era of advanced automation integration (Sarter, Mumaw, & Wickens, 2007). Continuing with the topic of eyes and vision, a new piece of technology that is catching the interest of human factors professionals are virtual reality headsets. These devices can unlock an entire new field of research that enlists the use of these devices for training purposes. Researchers are beginning to experiment with understanding the performance benefits or potential increases in performance metrics using virtual reality headsets in training. This training can be implemented in numerous fields, from firearm training and security to skilled laboring in industries to aviation.

Although we have enumerated some of the interesting new tools and devices being used in research, it is important to also mention some of the statistical analyses that are associated with such research. Several types of analyses go along with research methodologies using experimental designs, but two of the most interesting ones that we will focus on are mediation and factor analysis. Within mediation itself, there are several ways to achieve the results of mediation. The most cutting-edge method is detailed by Hayes (2013). Conversely, factor analysis has been around for many years; it is still one of the most popular statistical analyses. Several authors detail the methodology of conducting factor analysis (see Brown, 2014; O'Rourke & Hatcher, 2013). Principal component analysis and factor analysis have been commonly used in recent years for creating valid scales of measurement, which in turn can be used for several experimental research studies (Mehta & Rice, 2016; Mehta, Rice, Carstens, Cremer, & Oyman, 2015).

## APPLICATION

The ability to integrate technology and automation into almost every product, process, or system is probably one of the greatest assets of today's society (Taylor, 2006; Zaninotto & Plebani, 2010). It would be impossible to list or discuss all the current and potential applications of automation in the world today, but nonetheless, some standout winners deserve attention. The two main criteria for automation application within a system are to reduce workload and increase efficiency (Olson & Lucas, 1982). Earlier, automation applications were limited by the large financial cost of advanced technologies. However, in recent years, with more widespread research and development across several industries and more global implementations, we have seen the introduction of more automation due to drastic increases in economic viability (Bloss, 2013).

Before we discuss the general topic area of automation implementation, we will discuss a particular line of research that involves a compilation of items discussed in previous sections, the methodology, the statistical analyses, and the future implementation of automation in aviation. Rice et al. (2014) conducted a study on passenger perceptions of automation implementation in future airline cockpits. The study sought to determine consumer trust, comfort, and willingness to fly based on the configuration of the cockpit. The study utilized 104 participants from the United States and 97 participants from India, all collected from Amazon's Mechanical Turk. The participants were presented with a scenario of a commercial airline flight and told that the flight would be piloted with a human pilot, an autopilot, or a human pilot on the ground using a remote control system. They were then asked to rate their trust, willingness, and comfort on a 7-point Likert type scale from *extremely uncomfortable/distrust/unwilling* (–3) to *extremely comfortable/trust/willing* (+3) with a neutral position (0), based on whether they were flying, their colleague was flying, or whether their child was flying.

The results shown in Figure 6.1 indicate a statistically significant decrease in comfort, trust, and willingness based on the type of entity flying the aircraft. The study suggested that participants were less comfortable, less trusting, and less willing to fly on board the flights involving the completely automated aircraft and remote-control-operated aircraft compared to the human pilot in the cockpit.

Another study in this line of research was conducted by Mehta, Rice, Winter, and Oyman (2014). The study was completed with 194 participants from India and the United States using Amazon's Mechanical Turk. In this study, the participants were presented with a scenario of a commercial airline flight and told that the flight would be piloted with a human pilot, two human pilots on the ground using a remote control system, or a hybrid where one pilot was in the cockpit and one pilot was on the ground with remote control capabilities. They were then asked to rate their trust, willingness, and comfort on a 7-point Likert type scale from *extremely uncomfortable/distrust/unwilling* (–3) to *extremely comfortable/trust/willing* (+3), with a neutral position (0).

The results shown in Figure 6.2 indicate a statistically significant decrease in comfort, trust, and willingness based on the type of entity flying the aircraft, but the interesting aspect is that unlike the previous study, even though ratings toward the hybrid cockpit were lower, they were not negative, but rather neutral or slightly above neutral. This is an encouraging sign for the application of more automated cockpits. These studies put forth an interesting paradigm of analyzing the consumer's mindset regarding the use of automation within the aviation industry. It is important to note that while technology may allow for increased integration of automation, it may not always be the most practical or accepted option. Moving forward, it is important for human factors professionals to understand the consumer's point of view, even though automation is quite widespread and integrated into many daily tasks.

Automation is all around us. It is in our cars, in our homes, and even in our pockets. Over a billion cars are on the road worldwide, and that number is only increasing. A car is one of the most common possessions universally, and so the implementation of automation and the future of automobiles are of widespread significance to us all. Over the years, cars have evolved, in design, efficiency, power, aesthetics, and, of course, automation integration (Comacchio, Volpato, & Camuffo, 2012). Cruise control has been seen in cars as early as the late 1950s but has steadily become more universal since then. Today, it would be virtually impossible to find a

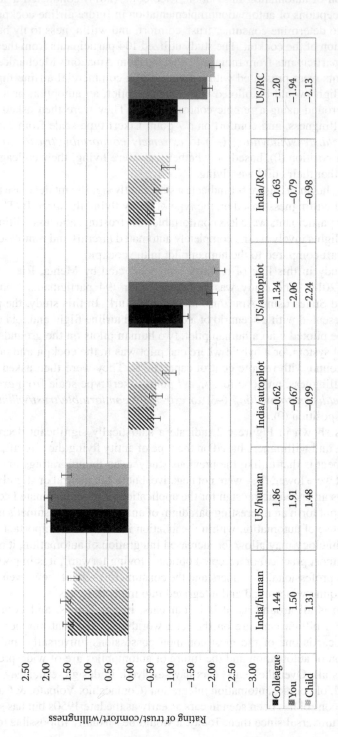

**FIGURE 6.1** Ratings of trust/comfort/willingness based on automation involvement and passenger type.

|  | India/human | US/human | India/autopilot | US/autopilot | India/RC | US/RC |
|---|---|---|---|---|---|---|
| Colleague | 1.44 | 1.86 | -0.62 | -1.34 | -0.63 | -1.20 |
| You | 1.50 | 1.91 | -0.67 | -2.06 | -0.79 | -1.94 |
| Child | 1.31 | 1.48 | -0.99 | -2.24 | -0.98 | -2.13 |

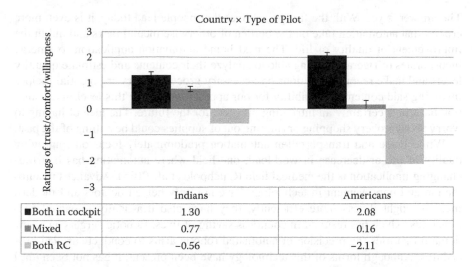

| Country × Type of Pilot | Indians | Americans |
| --- | --- | --- |
| ■Both in cockpit | 1.30 | 2.08 |
| ■Mixed | 0.77 | 0.16 |
| ■Both RC | −0.56 | −2.11 |

**FIGURE 6.2**   Ratings of trust/comfort/willingness based on automation involvement.

new car that does not integrate this level of automation into its repertoire of features. Although cruise control may be a standard automation feature in vehicles today, we are on the brink of the next level of this line of automation integration.

More advanced cars being produced today have several new levels of automation that can be incorporated, from active lane assist, to autonomous braking, and even self-driving "autopilot" capabilities. These new applications may not be as widespread yet, but they are certainly gaining traction, and it is not unfathomable to expect them to be standard features in the next 20 years (Litman, 2014). Leading manufacturers have already begun offering autonomous braking and accident avoidance systems as standard features on most of their models. The technological capabilities of the automation certainly seem up to the mark, but they rejuvenate a topic of discussion that has historically gone hand in hand with automation: Does increased automation reduce the operator's situational awareness? The argument has been around for years in the context of pilot overreliance on automation and increased complacency in task completion. Could this be the case now that cars are being fitted with comparable levels of automation?

The car is not the only aspect of our everyday personal lives that is seeing an upgrade with automation. If you look around your house, you can identify a couple of new automated features that have begun to creep in. Nest Labs is a company that has made a major emergence in to the home automation market (Withanage, Ashok, Yuen, & Otto, 2014). Homes can now be equipped with "smart" devices that are programmable, self-learning, sensor driven, and Wi-Fi enabled. The functionality of these devices can range from climate control to security. The idea of a self-learning device for indoor climate control has several benefits, not the least of which is energy saving. However, with the connectivity and communication we have these days, another great advantage is that our personal devices and smart phones can communicate with these home automation systems no matter where we are physically. Communication is one aspect of home automation, but could foresight be another?

The answer is yes. With the fast-paced lives that people lead today, it is even more crucial that automation take on the burden of otherwise menial tasks and aid in the improvement of quality of life. The next home automation application to emerge incorporates refrigerators being able to analyze their contents and estimate quantity levels and make recommendations on necessary grocery orders or potential recipes involving said contents. The ability for our appliances to have this level of automation built in is certainly an intriguing prospect for the future. The idea of having to worry about grocery shopping or running out of supplies could be a thing of the past.

While home and transportation automation predominately focus on improving quality of life or decrease in workload, one field where automation has had life-changing implication is the medical field (Catchpole et al., 2015). Advances in automation and improvement in technology have not only helped doctors and modern medicine fight illness more efficiently, they have also drastically impacted success rates, which has resulted in countless saving of lives. Robotic surgery is a new domain that uses the precision of automated robotic arms to conduct surgical procedures. Although forms of this technology have been present, it has not been until recent years that we have seen the applicability and use of such techniques become more widespread. Automation use through robotic surgery can increase the precision of the surgeon as well as drastically reduce the length of time taken to perform such surgical procedures (Lanfranco, Castellanos, Desai, & Meyers, 2004). Image guidance and augmented reality have aided minimally invasive surgical procedures and have the potential to increase safety, reduce errors, and more effectively deal with difficulties as they arise during surgery (Diana & Marescaux, 2015).

The examples discussed represent only some of the numerous applications of today's automation but nevertheless paint a vivid picture of the current state of affairs. Although it seems quite extraordinary, we must try to appreciate the fact that automation is quite literally in its infancy and there are plenty more advancements waiting around the corner. It is certainly an exciting prospect and only time will tell the level of automation we achieve in the decades to come.

## FUTURE TRENDS

The future for automation growth looks incredibly bright, which in turn has the potential to impact our daily lives in incomprehensible ways. We can only speculate how different the future of the world will look with new automated advancements, but if past experience is any predictor, the prospects are destined to be intriguing. Experts predict that we have not yet seen the explosion in new technology and automation yet but rather are on the cusp still of exponential growth in this field (Kurzweil, 2014). It is quite certainly an exciting time to be alive.

It seems like every day we are being introduced to new products and innovations that change the way we interact with the world around us. The nature of our interactions with our homes and our automobiles has already begun to evolve with the addition of new automated technologies. An entire new industry has been spawned in recent years with several companies redesigning and updating existing consumer products with new technologically advanced features and thereby changing the entire manner in which they are utilized (Tanner, 2015). For example, sunglasses have been around for decades, and their primary functionalities have been shielding eyesight and being fashion accessories. However, in recent years, they have received a tech upgrade. New companies now have introduced sunglasses with Bluetooth speakers built in. A sensor in the stem of the sunglasses picks up the person's jawbone frequency, allowing the device to be used to make phone calls as well (Brusie et al., 2015). This represents not necessarily a revolutionary technology, in either sense, but the marriage of these two products is the result of the advances we are beginning to see in automation applications. Therefore, it is a popular belief that there will be two paths we are going to see in terms of automation advancements in the future. One is the expected development of revolutionary new ideas that have never existed before, and the other, but equally interesting, will be that of evolutionary advancements represented by the concept of marrying two technologies or products that historically have never been considered to be one product.

The most salient of all associations with the term *automation* comes to us from the aviation industry. Aircraft have been at the forefront of automation implementation. The integration of automated aids such as autopilots, Global Positioning Systems, and Traffic Collision Avoidance Systems has all been seen for decades, and the field is still expanding (Hancock et al., 2013). We are on the verge of seeing the next big revelation in terms of aviation automation. Airline cockpits could be in for a drastic overhaul, with future technologies and designs already rumored to be in conception. Pilotless or at least single-pilot cockpits are almost certainly going to be part and parcel of the future of commercial air travel due to the economic advantages as well as due to the current shortage of pilots (Malik & Gollnick, 2016). Aircraft manufacturers will have to keep consumer perceptions in mind, as initial apprehension from passengers is more than likely to be expected with such drastic changes. Autonomous cockpits, however, represent only one aspect of the future of aviation. Aviation experts have also begun discussing the viability of suborbital space flight with advanced automated aircraft being able to complete transoceanic flights in a fraction of the time that current airliners take to achieve the same (Webber, 2013a). Speaking of destinations, the advancements made by public and private space travel agencies and companies will most certainly trickle down to commercial air travel and unlock the potential for space tourism (Webber, 2013b). Could we be one day booking trips to the moon as part of our vacation plans?

Another avenue of future growth for the automation industry looks to come from mimicking the trends in aviation. Every automotive giant in the world has had to take note of newcomers Tesla and acknowledge the fact that the game is changing. Tesla has effectively thrown down the gauntlet to other manufacturers, and we are about to witness the biggest evolutions in car design and manufacturing since the invention of the automobile. Are driverless cars going to be the accepted norm in 50 years (Litman, 2014)? It certainly seems that we are on a path to such a future. Will the automation be robust and safe enough to handle the biggest traffic jams of lesser developed countries with less than optimal road regulations? Or is this is just a fad that will eventually fade into a novelty sector of the automotive market? The ethical dilemmas associated with driverless cars also need serious attention if this is going to be a viable pathway for the future (Duffy & Hopkins, 2013; Greene, 2016). Circling back to the concept of marrying two forms of technology, in recent years, innovators have also delivered a functional prototype for flying cars or, more specifically, cars that can transform into airworthy flying machines. The most promising of these designs comes to us from AeroMobil (Marks, 2014; Rajashekara, Wang, & Matsuse, 2016).

We would be remiss to not discuss future advancements in automation within the medical field, as it potentially is the most essential to longevity of life. Over time, the advancements in modern medicine and science have directly resulted in human beings being able to live longer and healthier lives. It is therefore quite certain that this trend will continue with the assistance of automation in medicine today. Automation could soon make its breakthrough into the medical schools with new virtual and augmented reality devices and virtual anatomy tables allowing students to hone their skills and practice complex medical procedures (Fang, Wang, Liu, Su, & Yeh, 2014; Paech et al., 2016). Medication delivery methods could also be seeing new innovations with the use of automation for more accurate, concentrated, and targeted delivery of drugs in order to increase effectiveness, reduce time to take effect, as well as potentially reduce side effects (Woods & Constandinou, 2013). Automation as a whole is most definitely going to see us dealing with medical issues more effectively and help us deal with illness much better than we have been able to in the past.

## CONCLUSION

Automation is likely to be one of the major technological platforms that will bridge the gap from where we are today to our future. The manner in which our understanding and development of automation take shape will most certainly play a role in what our daily lives will look like in the years to come. The challenge we face now is accurately applying our current knowledge on automation in new and innovative ways to improve our future. Over the last couple of decades, we have gone from having a very rudimentary understanding of the relationship between humans and automation to very sophisticated and complex models of said relationship. Connecting where we were to where we are to where we are going is the key to successful automation implementation in the future. While the main aim of this chapter is to give a cursory outline of the fundamentals of automation and its corresponding research, the body of knowledge on automation is immensely vast and cannot be done justice in a single chapter. The most salient and fundamental aspects of the automation literature have been discussed with an eye toward exploring the future applications of such technologies and theories.

## REFERENCES

Bailey, N. R., & Scerbo, M. W. (2007). Automation-induced complacency for monitoring highly reliable systems: The role of task complexity, system experience, and operator trust. *Theoretical Issues in Ergonomics Science*, 8(4), 321–348.

Billinghurst, M., Clark, A., & Lee, G. (2015). A survey of augmented reality. *Foundations and Trends Human–Computer Interaction*, 8(2–3), 73–272.

Bloss, R. (2013). Innovations like two arms, cheaper prices, easier programming, autonomous and collaborative operation are driving automation deployment in manufacturing and elsewhere. *Assembly Automation*, 33(4), 312–316.

Bolton, M. L., & Bass, E. J. (2010). Formally verifying human–automation interaction as part of a system model: Limitations and tradeoffs. *Innovations in Systems and Software Engineering, 6*(3), 219–231.

Brown, T. A. (2014). *Confirmatory factor analysis for applied research.* New York: Guilford Publications.

Brusie, T., Fijal, T., Keller, A., Lauff, C., Barker, K., Schwinck, J., Calland, J. F., & Guerlain, S. (2015). Usability evaluation of two smart glass systems. In *Systems and Information Engineering Design Symposium (SIEDS), 2015* (pp. 336–341). Charlottesville, VA: IEEE.

Byrne, E. A., & Parasuraman, R. (1996). Psychophysiology and adaptive automation. *Biological Psychology, 42*(3), 249–268.

Catchpole, K., Perkins, C., Bresee, C., Solnik, M. J., Sherman, B., Fritch, J., Gross, B., Jagannathan, S., Hakami-Majd, N., Avenido, R., & Anger, J. T. (2015). Safety, efficiency and learning curves in robotic surgery: A human factors analysis. *Surgical Endoscopy, 30* (9), 3749–3761.

Chow, S., Yortsos, S., & Meshkati, N. (2014). Asiana Airlines Flight 214. *Aviation Psychology and Applied Human Factors, 4*, 113–121.

Comacchio, A., Volpato, G., & Camuffo, A. (Eds.). (2012). *Automation in automotive industries: Recent developments.* New York: Springer Science & Business Media.

Diana, M., & Marescaux, J. (2015). Robotic surgery. *British Journal of Surgery, 102*(2), e15–e28.

Duffy, S., & Hopkins, J. P. (2013). Sit, stay, drive: The future of autonomous car liability. *Social Science Research Network.* Retrieved from http://papers.ssrn.com/sol3/papers.cfm?abstract_id=2379697

Endsley, M. R. (1996). Automation and situation awareness. In R. E. Parasuraman & M. E. Mouloua (Eds.), *Automation and human performance: Theory and applications* (pp. 163–181). Mahwah, NJ: Lawrence Erlbaum.

Endsley, M. R. (1999). Level of automation effects on performance, situation awareness and workload in a dynamic control task. *Ergonomics, 42*(3), 462–492.

Fang, T. Y., Wang, P. C., Liu, C. H., Su, M. C., & Yeh, S. C. (2014). Evaluation of a haptics-based virtual reality temporal bone simulator for anatomy and surgery training. *Computer Methods and Programs in Biomedicine, 113*(2), 674–681.

Greene, J. D. (2016). Our driverless dilemma. *Science, 352*(6293), 1514–1515.

Hancock, P. A., Jagacinski, R. J., Parasuraman, R., Wickens, C. D., Wilson, G. F., & Kaber, D. B. (2013). Human–automation interaction research past, present, and future. *Ergonomics in Design, 21*(2), 9–14.

Hayes, A. F. (2013). *Introduction to mediation, moderation, and conditional process analysis: A regression-based approach.* New York: Guilford Press.

Kurzweil, R. (2014). The singularity is near. In R. L. Sadler (Ed.), *Ethics and emerging technologies* (pp. 393–406). Basingstoke. UK: Palgrave Macmillan.

Lanfranco, A. R., Castellanos, A. E., Desai, J. P., & Meyers, W. C. (2004). Robotic surgery: A current perspective. *Annals of Surgery, 239*(1), 14–21.

Lee, J. D., & See, K. A. (2004). Trust in automation: Designing for appropriate reliance. *Human Factors, 46*(1), 50–80.

Litman, T. (2014). Autonomous vehicle implementation predictions. *Victoria Transport Policy Institute, 28.*

MacKay, K., & Vogt, C. (2012). Information technology in everyday and vacation contexts. *Annals of Tourism Research, 39*(3), 1380–1401.

Malik, A., & Gollnick, V. (2016). Impact of Reduced Crew Operations on Airlines-Operational Challenges and Cost Benefits. In *16th AIAA Aviation Technology, Integration, and Operations Conference* (p. 3303). Washington, DC: AIAA.

Marks, P. (2014). Hop in, I'm flying. *New Scientist, 224*(2992), 19–20.

Mehta, R., & Rice, S. (2016). Creating a short scale for in-flight experience quality. *International Journal of Social Science and Humanity, 6*(12), 970.

Mehta, R., Rice, S., Carstens, D., Cremer, I., & Oyman, K. (2015). A brief Intermodal Rail Network (IRN) scale: Establishing validity and reliability. *Journal of Sustainable Development, 8*(6), 243.

Mehta, R., Rice, S., Winter, S. R., & Oyman, K. (2014, October). Consumers' perceptions about autopilots and remote-controlled commercial aircraft. *Proceedings of the 57th Annual Meeting of the Human Factors and Ergonomics Society.* Chicago, IL: HFES.

Merritt, S. M., & Ilgen, D. R. (2008). Not all trust is created equal: Dispositional and history-based trust in human automation interactions. *Human Factors, 50*(2), 194–210.

Meyer, J. (2001). Effects of warning validity and proximity on responses to warnings. *Human Factors, 43*, 563–572.

Olson, M. H., & Lucas Jr., H. C. (1982). The impact of office automation on the organization: Some implications for research and practice. *Communications of the ACM, 25*(11), 838–847.

Onnasch, L., Wickens, C. D., Li, H., & Manzey, D. (2013). Human performance consequences of stages and levels of automation: An integrated meta-analysis. *Human Factors, 56*(3), 476–488.

O'Rourke, N., & Hatcher, L. (2013). *A step-by-step approach to using SAS for factor analysis and structural equation modeling.* Cary, NC: SAS Institute.

Paech, D., Giesel, F. L., Unterhinninghofen, R., Schlemmer, H. P., Kuner, T., & Doll, S. (2016). Cadaver-specific CT scans visualized at the dissection table combined with virtual dissection tables improve learning performance in general gross anatomy. *European Radiology, 27*(5)1–8.

Parasuraman, R., & Riley, V. (1997). Humans and automation: Use, misuse, disuse, abuse. *Human Factors, 39*(2), 230–253.

Parasuraman, R., Sheridan, T. B., & Wickens, C. D. (2000). A model for types and levels of human interaction with automation. *IEEE Transactions on Systems, Man, and Cybernetics—Part A: Systems and Humans, 30*(3), 286–297.

Rajashekara, K., Wang, Q., & Matsuse, K. (2016). Flying cars: Challenges and propulsion strategies. *IEEE Electrification Magazine, 4*(1), 46–57.

Rice, S. (2009). Examining single- and multiple-process theories of trust in automation. *The Journal of General Psychology, 136*(3), 303–322.

Rice, S., Kraemer, K., Winter, S. R., Mehta, R., Dunbar, V., Rosser, T. G., & Moore, J. C. (2014). Passengers from India and the United States have differential opinions about autonomous auto-pilots for commercial flights. *International Journal of Aviation, Aeronautics, and Aerospace, 1*(1), 3.

Sarter, N. B., Mumaw, R. J., & Wickens, C. D. (2007). Pilots' monitoring strategies and performance on automated flight decks: An empirical study combining behavioral and eye-tracking data. *Human Factors, 49*(3), 347–357.

Scerbo, M. W. (1996). Theoretical perspectives on adaptive automation. In R. E. Parasuraman & M. E. Mouloua (Eds.), *Automation and human performance: Theory and applications* (pp. 37–63). Mahwah, NJ: Lawrence Erlbaum.

Sheridan, T. B., & Verplank, W. L. (1978). *Human and computer control of undersea teleoperators.* Cambridge, MA: Massachusetts Inst. of Tech. Cambridge Man-Machine Systems Lab.

Tanner, A. N. (2015). The emergence of new technology-based industries: The case of fuel cells and its technological relatedness to regional knowledge bases. *Journal of Economic Geography, 16*(3), 611–635.

Taylor, R. M. (2006). *Human automation integration for supervisory control of UAVs.* Farnborough, UK: Defence Science and Technology Lab.

Webber, D. (2013a). Point-to-point people with purpose—Exploring the possibility of a commercial traveler market for point-to-point suborbital space transportation. *Acta Astronautica, 92*(2), 193–198.

Webber, D. (2013b). Space tourism: Its history, future and importance. *Acta—Astronautica*, *92*(2), 138–143.

Wickens, C. D., (1992). *Engineering psychology and human performance* (2nd ed.). New York: Harper Collins.

Wickens, C. D., & Hollands, J. G. (2000). *Engineering psychology and human performance* (3rd ed.). Upper Saddle River, NJ: Pearson Prentice Hall.

Withanage, C., Ashok, R., Yuen, C., & Otto, K. (2014, May). A comparison of the popular home automation technologies. In *2014 IEEE Innovative Smart Grid Technologies— Asia (ISGT ASIA)* (pp. 600–605). Kuala Lumpur, Malaysia: IEEE.

Woods, S. P., & Constandinou, T. G. (2013). Wireless capsule endoscope for targeted drug delivery: Mechanics and design considerations. *IEEE Transactions on Biomedical Engineering*, *60*(4), 945–953.

Zaninotto, M., & Plebani, M. (2010). The "hospital central laboratory": Automation, integration and clinical usefulness. *Clinical Chemistry and Laboratory Medicine*, *48*(7), 911–917.

## KEY TERMS

**automation:** completion of a task or some work by a machine or device that would otherwise need to be completed by a human (Wickens & Hollands, 2000).

**autopilot:** use of automated aids and devices to help a pilot control the aircraft and perform basic flight maneuverings.

**experimental design:** use of a particular type/form of setting up a research experiment involving the manipulation of a variable to identify an effect on another variable, establishing a cause–effect relationship.

**factor analysis:** another type of statistical analysis that helps group correlated variables into smaller groups based on common features or functionality.

**mediation:** type of statistical analysis that suggests that when a relationship between an IV and a DV is explained by a third variable, this variable is said to be the mediator variable and is said to have a mediating/indirect effect on the relationship. This process is known as mediation.

**methods:** refers to the different way and means of performing research studies in order to answer different research questions.

**statistical analyses:** refers to different mathematical calculations and ways of measuring data collected during research experiments in order to help answer the research questions (e.g., analysis of variance, mediation).

# 7 User-Centered Design in Practice

*Raegan M. Hoeft and Debbie Ashmore*

## CONTENTS

## INTRODUCTION

With the continuing advances in and our ever-increasing reliance on technology in the last 30 years, it should come as no surprise that efforts to make interactive technologies easier to use have become more important to everyone. Technology has become ubiquitous throughout our daily lives, from transportation to work to communication to healthcare to entertainment. At the same time, we can all recall at least one disastrous experience with technology where we were left feeling confused, frustrated, and wondering why we could not figure out how to perform some seemingly easy task. While significant strides have been made in the user-friendliness of some kinds of technology (e.g., consumer products), room for improvement is still warranted (e.g., enterprise software for employees). This becomes even more obvious as our expectations are shaped by the cutting-edge, novel technologies we are introduced to on a regular basis.

In the mid-1980s, when it became clear personal computers were not simply a passing fad, the notion that developers and designers needed to focus more attention on the end user of interactive technologies started to take shape. One of the first references to this user need was when Gould and Lewis (1985) published an article in which they advocated for the addition of three design concepts, which were not a standard part of system design: (a) an early focus on and interactions with end users, (b) an iterative design process, and (c) empirical measurement. Norman (1986) expanded on these concepts by coining the term *user-centered design* (UCD). He stated designs should "start with the needs of the users [and to the user] the interface is the system" (p. 61); this statement has had a significant impact on how we think about the user experience (UX). Since these initial publications, UCD has broadened and evolved exponentially, influencing how we approach the design, development, and evaluation of technology, as well as how we attempt to understand and speak for the different kinds of end users that exist.

Yet, the concept of UCD is still not well understood by many outside of the human factors and related communities and, consequently, is not always supported or valued by the development team. In fact, although to some, UCD seems like common sense, to others, it can feel like a waste of time and money, something that magically happens or something anyone can do without any formal training. These are often the people who believe it is the user's fault if the system is not used as designed, while others believe any user-centric issues can be mitigated by documentation, training, or maintenance and support (see Figure 7.1; Kjær-Hansen, 1999). (We cringe every time we hear this!) Due to this pass-the-buck mentality, the world is still plagued with poorly designed interfaces, which are often "engineering" or "software development" displays that are designed by the writers of the code and are not understood by the average user of the system. This unfortunate reality is not only frustrating but

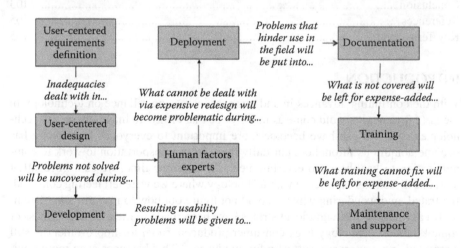

**FIGURE 7.1**   Passing the buck. (Adapted from Kjær-Hansen, 1999. *A business case for human factors investment.* European Organisation for the Safety of Air Navigation. Report HUM. ET1.ST13.4000-REP-02. Retrieved from http://www.idemployee.id.tue.nl/g.w.m.rauterberg /lecturenotes/HF99cost-report.pdf.)

can have costly (e.g., increased development costs, extensive redesign, and longer time to market; Wilson & Rosenbaum, 2005) and/or tragic consequences, as evidenced by stories in Steven Casey's *Set Phasers on Stun* (1998) and *The Atomic Chef* (2006), both often used in introductory Human Factors courses.

## FUNDAMENTALS

The concept of UCD may have broadened, but the essential principles remain the same: focus design efforts on the users and include users in the process to the extent possible. Three main components, or project phases, of UCD are (a) understanding your users (user research), (b) creating and refining a design for the user interface (iterative design), and (c) ensuring what you have designed is accomplishing the goals of the design (evaluation). In addition, there is a critical phase beyond UCD where UCD practitioners can be instrumental advocates for the users during development and deployment (see Figure 7.2).

In this chapter, we will specifically focus on UCD related to software systems and graphical user interfaces (GUIs); however, many of these principles apply to UCD in the context of other types of interfaces as well (such as physical workstations or amusement park attractions).

A number of reasons why one might be engaging in UCD related to software systems include, but are not limited to: (a) a legacy system is being redesigned, refreshed, or enhanced, (b) a new or missing component is being incorporated into a legacy system, (c) a new tool/idea is shopping for a home/user group, (d) a new system is being designed from scratch (no legacy system), and/or (e) a new user group is being introduced for an existing system. It is always important to understand the underlying project goal(s) first so the UCD approach and methods can be scoped to align with the expected and anticipated outcomes of the entire effort. Although the ultimate goal of UCD is clearly to improve the UX, the *user research*, *iterative design*, and *evaluation* activities should all be selected with the driving force behind the project in mind. If UCD professionals take the high ground and resist compromise, they may be quickly ushered to the door. We will now discuss the fundamentals of each of the phases and how they all fit together to create a full UCD process.

### USER RESEARCH

The goal of *user research* is to develop a solid understanding of the design challenge or problem space to guide the forthcoming design process(es). User research encompasses all the activities conducted to understand end user(s), including who they are, what their goals are, what their preferences are, how they currently complete their tasks, what their pain points are, etc. It can be a blend of a variety of different

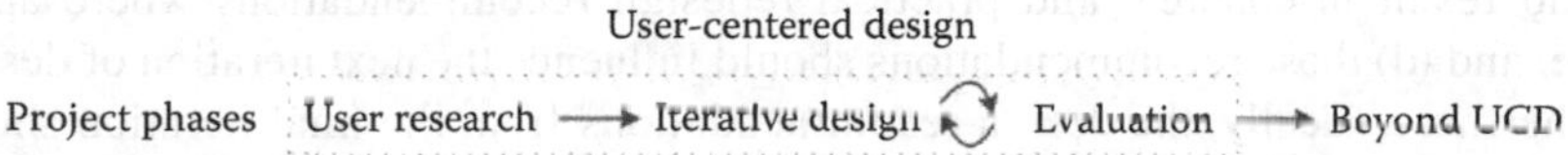

User-centered design

Project phases : User research ⟶ Iterative design ↻ Evaluation ⟶ Beyond UCD

**FIGURE 7.2** The UCD process project phases.

kinds of research, including field research, ethnography, ergonomic analysis, and knowledge elicitation activities, among other things. These tools are used to extract information from users and environments in various ways and apply a critical eye to interpret the findings in light of the problem space. Outcomes from user research include workflow and/or task analysis diagrams, user profiles, personas, scenarios, customer journey maps, and experience maps. User research findings are also used to compose the user-centered requirements that drive iterative design in the next phase of UCD. User research findings are the most reusable component of UCD because users often use multiple systems, have many tasks, represent part of a company culture, etc. An understanding of the characteristics of users and their environments is valuable beyond their application to a specific project.

## ITERATIVE DESIGN

Iterative design includes all the activities related to the actual design of the system, with a focus on the user interface component of the system. The broad responsibility of a UCD designer is to ensure that a product or system is designed in such a way that it logically flows from one interaction step to the next. The goal of *iterative design* is thus to create the user interface design and product flow (literally draw the layout, components, content, storyboards, sitemaps, etc.) through a series of increasingly more detailed and complete attempts (i.e., low-fidelity sketches to high-fidelity, interactive prototypes), until a design emerges that meets some threshold of acceptance specified by your target customer and your own team. Ideally, designs should emerge from the output of user research, basic human–computer interaction and display principles, and one's past experiences with similar problem spaces, all blended with representative user feedback in each iteration to take the designs closer to a final acceptable state. The outcomes from iterative design include information architectures, content analysis, detailed interaction designs, wireframes, mockups and interactive prototypes, and visual and functional designs, which are ready for evaluation and user testing.

## EVALUATION

With the intention to gather user feedback in each iteration, the third phase of UCD is evaluation. The goal of *evaluation* is to understand the strengths and weaknesses of designs, in other words, to find out what is working well for users and what is not. Evaluation methods vary and can include both expert-directed assessments (e.g., heuristic evaluations, compliance reviews) and user-based assessments (e.g., informal user feedback sessions, usability testing, human performance testing). Ideally, (a) some form of evaluation should accompany each design iteration, starting with evaluation of legacy, current, and/or competitive systems, (b) evaluations should include end user participation (to the extent possible and practical), (c) they should result in concrete and practical redesign recommendations where applicable, and (d) those recommendations should influence the next iteration of design. Furthermore, ideally, the design recommendations from the final iteration should be incorporated into updated designs prior to development. Due to myriad reasons, this is not always the case and the evaluation reports often serve as documentation for what should be done and what may lead to decrements in the UX after deployment. The outcomes from evaluation include practical design recommendations with each design iteration, metrics to compare design iterations to baseline (legacy, current, and/or competitive systems), and formal usability testing results that support finalizing designs.

## Beyond UCD

Finally, once designs are evaluated, the attention shifts to the activities *beyond UCD*, which consist of development, deployment, and other activities related to getting a system/product ready to be used by its intended users in the field. During development, the role of UCD practitioners is to collaborate with the development team to ensure that designs are implemented in a way that best meets user needs, to help interpret the user-centered requirements and designs during design reviews, and to work through implementation challenges that may require design tweaking. The UCD activities and artifacts from early phases will be rendered inconsequential if they are not understood or implemented in accordance with users' needs. While there will always be tradeoffs with regard to design and technology decisions, it is important to have a UCD practitioner continuously involved during development to help prioritize and advocate for users. To support deployment, results from the UCD evaluation phase can help guide the design of necessary documentation, training materials, and maintenance and support materials (see Figure 7.1).

## Summary

While there are three UCD project phases and one beyond the UCD phase, in many cases, this process may be conducted simultaneously yet staggered across multiple features of a system/product development lifecycle, in that some features are in development while others are still in the design phase and others are in the user research phase. This is especially true for agile software development, which focuses on delivering different features in sprints, generally between two and four weeks long. In contrast, in traditional waterfall systems engineering, the requirements and designs are supposed to be locked in prior to the development phase beginning (c.f., Eberts, 1994; Palmquist, Lapham, Miller, Chick, & Ozkaya, 2013). While iterative design is not prohibited in classic waterfall development efforts, any changes in design involve costly rework as each phase concludes with a formal review and extensive documentation and thus UCD oversights are expensive to correct when they are uncovered in later stages (Mohammed, Munassar, & Govardhan, 2010).

In practice, the UCD process is rarely conducted in full; shortcuts and compromises are common within each phase and across the phases. Various types of constraints require the UCD process and phases to be tweaked to accommodate those constraints; the goal of UCD practitioners is to attempt to do as much as possible to identify, understand, and represent users' needs given these constraints, thereby maintaining some level of user influence over the development process. Table 7.1 provides details about ideal characteristics of each UCD phase in comparison to the realistic conditions that often have to be contended with.

## TABLE 7.1
## UCD Process: Ideal Conditions vs. Realistic Conditions

| Phase | Ideal Conditions | Realistic Conditions |
|---|---|---|
| User research | –Time available to fully understand all user groups, the current system, and business needs<br>–Time to document user research for future use (reusability)<br>–Baseline metrics gathered on existing system for later comparison<br>–Unbiased view into what users really need<br>–Findings analyzed and interpreted by qualified UCD researcher(s) | –No time for research, must start design immediately<br>–Client already "knows" what users need and want and will dictate requirements without UCD research<br>–Access to stakeholders (SMEs, managers, etc.) but not actual target end users<br>–Client disagrees with user research results and makes decisions based on his/her needs/wants<br>–Stakeholders have an agenda already and only want research that supports that agenda (want validation of beliefs/ desires, not empirical findings)<br>–Requirements are fluid and changing as the project is happening |
| Iterative design | –Designs are primarily based on the needs and wants of end users<br>–Begin with low-fidelity sketches, iterate based on user feedback, eventually get to high fidelity prototypes<br>–Begin with high-level UX concepts and then move toward low-level features and associated detailed designs<br>–Users understand designs are malleable and do not actually exist yet | –Designs must conform to existing style guides or pattern libraries<br>–No time for multiple iterations; must get first design as good as possible<br>–Users get distracted easily by incorrect or erroneous design elements<br>–Development team/others have been doing their own designs for years and question the need for the support of UCD practitioners<br>–Development team does not want UCD inputs and is openly hostile or simply ignoring UCD considerations<br>–Development team just wants UCD practitioners to quickly provide wireframes and then move out of the way<br>–UCDs are deemed "technically unfeasible" (in truth or not)—too difficult to code<br>–UCD data are not available or in a format that supports requirements definition<br>–Focus is placed on very specific features and not the big picture (i.e., the overall UX)<br>–Belief that any design issues will be "fixed with training" or maintenance and support |

*(Continued)*

**TABLE 7.1 (CONTINUED)**
**UCD Process: Ideal Conditions vs. Realistic Conditions**

| Phase | Ideal Conditions | Realistic Conditions |
|---|---|---|
| Evaluation | –Heuristic evaluations of current or competitive systems up front<br>–Informal feedback and practical design recommendations with each design iteration<br>–Metrics to compare design iterations to baseline (current or competitive systems)<br>–Formal usability testing to wrap up design work | –Shortened timeframe for evaluation<br>–No access to target users to participate in evaluation(s)<br>–Scope reduced to accommodate changes in resources or schedule<br>–Client wants method (e.g., eye tracking) that does not match with goals of evaluation<br>–Stakeholders or other observers hijack user sessions<br>–Designs are already locked in, client just wants a check in the box and cannot actually make any UCD changes recommended at this point<br>–Client interprets findings based on literal, word-for-word statements made by individual users as opposed to prioritized summary of findings |
| Beyond UCD | –Document designs and research findings to maintain institutional knowledge about design decisions<br>–Collaborate with development team and provide UCD guidance to support design decisions during implementation<br>–Support design reviews<br>–Support creation of documentation, training materials, and maintenance and support materials | –Designs are handed off and development team directs implementation without further UCD practitioner support; final implemented designs often deviate from UCD designs<br>–Move on to the next effort before fully documenting project lessons-learned |

## METHODS

The methods associated with UCD are varied due to the vastly different types of problem spaces and domains, as well as the multidisciplinary nature of the practitioners who make up the field. Practitioners with different backgrounds have brought methods with them and these have merged and evolved over time such that the toolbox of UCD methods is overflowing and constantly expanding. Hundreds of different methods, activities, techniques, etc. have been identified for UCD across user research, iterative design, and evaluation and they can be combined and/or tailored to meet the goals and realistic conditions of any project. Figure 7.3 presents some of the more commonly used methods and activities as they apply to the project phases.

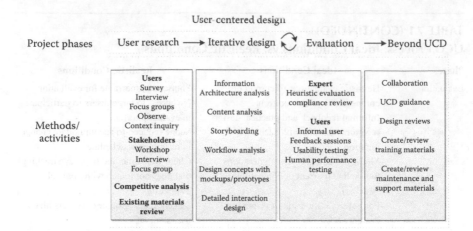

**FIGURE 7.3**  UCD commonly used methods/activities.

## User Research

Activities in the user research phase can be divided into three categories: (a) activities involving end users, (b) activities involving stakeholders, and (c) activities that focus on existing materials (e.g., interacting with an existing or legacy system, reading literature such as user manuals, taking training courses or reviewing training materials, etc.). Reviewing existing materials first is a great way to get educated on the system and/or domain of interest, learn some of the specific terminology, and get a firsthand glimpse into some of the existing problems. This is also a valuable way to prepare for interactions with end users and stakeholders; showing you have done your research in advance can lend you credibility and also facilitate richer questions and conversations. In some cases, unfortunately, access to end users and stakeholders is limited and the majority of your information must come solely from these existing materials.

As its name implies, interacting with end users is one of the main goals of UCD, so activities focused on interactions with end users should be prioritized. Contextual inquiries, in which end users are individually observed and interviewed in their standard work environment, provide the richest data about what users are actually doing and under what conditions. Wherever possible and practical, contextual inquiries should be conducted. Observations, even in the absence of speaking with end users, can yield significant data that are unbiased by the subjective opinions of end users. When interacting with end users, UCD practitioners generally prefer either one-on-one interviews or focus groups; however, in reality, we take what we can get and will conduct whichever is appropriate and/or realistic. Finally, surveys are useful when it may be difficult to get face time with end users due to proximity, timing, or any other reason. While surveys do not allow UCD practitioners to probe further into the answers provided, they can still be valuable in the absence of other data points.

In conjunction with research with end users, interactions with business stakeholders can yield information about other aspects of the problem space, including the sometimes very sensitive perspectives around existing technology (if applicable) and efforts to change to something new. The stakeholders will bring with them personal agendas based on their specific roles and priorities that may be in direct conflict with each other or with what comes out of user research. Interactions with stakeholders, through interviews, focus groups, or even creative workshops, can provide insight into a top–down perspective on end users' tasking, resources, and experiences, which may or may not align with the end users' perspectives on these things. These interactions are critical to understanding what will make your project successful from a business perspective, beyond creating a solution that meets your users' needs.

For a thorough description of user research and the various methods available, see the book of Courage and Baxter (2005), *Understanding Your Users,* or the chapter of Spath, Hermann, and Sproll's (2012), *User Requirements Collection and Analysis.*

## Iterative Design

Although design activities are often thought of as involving the design of mockups and prototyping, the other activities shown in Figure 7.3 are UCD iterative design activities that help flesh out ideas first, essentially fostering detailed planning before designing what end users will actually interact with. Information architecture and content analysis activities help a designer assess what information is or should be contained within a user interface and how it is or should be organized. Storyboarding and workflows, which consist of creating a visual path through a design's screens or planned functionality, allow a designer to lay out the order of how information will be presented or functionality will be organized into logical sequences. Interaction design activities focus on how the interface should behave to accomplish necessary goals and to align with the other design planning activities. These design activities all assist designers in developing user interface designs that are supportive of users, their information needs, and their goals and activities.

Prototyping a user interface can be done with paper and pencil (simple, low-fidelity prototypes) or using software prototyping tools like Axure, Invision, Balsamiq, or Microsoft Visio (fully interactive, high-fidelity prototypes), just to name a few. Each of the methods for prototyping has its pros and cons and a decision should be based on the project goals, constraints, and timeline. For example, if the project is in an early part of the iterative design UCD stage, and the goal is to understand if the design team is headed in the right direction, a low-fidelity prototype of the notional user interface is best because it is lightweight and can be discarded without regard to large amounts of wasted time and effort. Pros of low-fidelity prototypes include detecting potential usability problems at a very early stage in the design process before any code has been written, promotion of communication between developers and users, quick to build and refine, require little resources and materials to develop, and it is possible to show them to a wide audience via projector. The cons are based on their simplicity—it is not possible to evaluate details or measure information about usage. Therefore, low-fidelity prototypes are often used to test the design team's initial ideas so that the most useful and efficient design that meets the user's needs can emerge. On the other hand, if the project is in a later evaluation UCD stage, a low-fidelity prototype will not help the design team answer the types of questions they need to push through to the final design. As mentioned earlier, the outcomes from evaluation include practical design recommendations based on formal usability testing results. Therefore, it is best to budget and plan for a high-fidelity prototype in later design stages so appropriate user evaluation metrics (such as number and types of user errors while trying to complete a task) can be gathered while the design team observes actual users interacting with the prototyped user interface. A system that has been through prototyping will generally have an improved design quality and will be far closer to what the user needs.

Creating prototyped designs in an iterative fashion allows for flexibility in trying out multiple design ideas, comparing designs to each other, and focusing on different aspects of the designs at different iterations. Several rounds of iteration are desirable; however, too many iterations can lead to diminishing returns while too few iterations can result in less successful designs.

For a thorough description of iterative design, see the book of Stone, Jarrett, Woodroffe, and Minocha (2005), *User Interface Design and Evaluation*, and for a review of design principles, see the book of Lidwell, Holden, and Butler (2010), *Universal Principles of Design, Revised and Updated: 125 Ways to Enhance Usability, Influence Perception, Increase Appeal, Make Better Design Decisions, and Teach Through Design.*

## EVALUATION

Many different ways to evaluate a user interface design and associated UX are available, each with its own advantages and disadvantages, and each with its appropriate application. Evaluations can be differentiated on a number of factors, which include but are not limited to the following:

- Formative vs. summative
- Analytical vs. empirical
- Quantitative vs. qualitative
- Rapid vs. rigorous
- Expert vs. end user participation
- Focus on specific design aspects vs. open discovery
- Remote vs. in person
- Manual vs. automated

Evaluation methods can be tweaked as needed to meet the specific needs of each project. The methods should be selected based on the goal of the evaluation (often dictated by the iteration); however, more often than not, the evaluation method is selected based on time and resource constraints (specifically with regard to how many and what kind of end users can participate in an evaluation).

For a thorough description of evaluation, see the book of Stone et al. (2005), *User Interface Design and Evaluation*, and for a plethora of evaluation methods (and other research methods), see Martin and Hanington's (2012) *Universal Methods of Design: 100 Ways to Research Complex Problems, Develop Innovative Ideas, and Design Effective Solutions.*

## BEYOND UCD

The focus on the beyond UCD stage is mainly on collaborating with the development team by being present and available during development and deployment to ensure that the essence of what was created from the UCD process is maintained to the extent possible. Participating in meetings, such as design reviews, where the team makes decisions that may impact the design is an important activity for a UCD practitioner. Also, although there is likely documentation that was created during the project, this stage in the process is a good time to create any additional documentation that can be used for institutional knowledge down the line when there are questions about the who, what, where, when, why, and how of the user interface being designed the way it was. In addition, during this phase, UCD practitioners can also enhance the usability of training, maintenance, and other support documentation, ensuring they are written with their respective end users in mind.

## SUMMARY

We have presented the UCD process and methods and activities applicable to each of the phases. UCD practitioners continue to create new methods to tap into different aspects of the UX or to refresh traditional methods that may no longer be applicable to emerging technologies. Whichever methods and activities you select, you should do so with the goals of your project in mind. In this field, it is acceptable to use traditional methods, but it is also acceptable, and often necessary, to be flexible and creative in order to keep the user at the forefront throughout the development process.

## APPLICATION

We present now a case study for applying UCD to the user interface design for the Lockheed Martin Marlin Autonomous Underwater Vehicle (AUV). The Marlin AUV conducts subsurface surveys and inspections, often after severe weather events, to determine whether there has been any damage to oil and gas offshore structures and/or pipelines (see Figure 7.4). The Marlin inspects and generates real-time three-dimensional (3D) models of various fixed platforms; the goal of the user interface is

**FIGURE 7.4** The Marlin AUV. Copyright 2017 by the Lockheed Martin Corporation.

to aid the vehicle operator in the creation of mission plans, vehicle safety checkout, observation of the sortie, inspection of the 3D models, and postmission analysis. From a design perspective, the goal was to create a single display with which end users could successfully complete their mission, from preplanning to launch of the Marlin to conducting the actual inspection to recovery of the Marlin, and, finally, to postmission analysis.

Via stakeholder interviews and working group discussions, the team began the Marlin design project by defining four main goals:

1. create a unified system that consolidated the functionality of several legacy software systems,
2. create a contextualized display with all information optimized for task-centric user focus,
3. provide the functionality to complete the full mission, and
4. be single-operator capable.

The UCD steps that the design team took to approach the project encompassed the full process of user research, iterative design, and evaluation. In the user research phase, the team interacted with the end users by (a) observing the current test crew while they performed mock launch and recovery exercises dockside and (b) conducting interviews while walking through the legacy software systems. From these activities, the design team was able to determine what features and functionality the user interface would need. A card sorting exercise was conducted to organize those tasks into groups and determine the information architecture—this was the beginning of the iterative design phase. The card sorting exercise was conducted by taking what was learned during the user research phase and categorizing discrete tasks into mission phases. Each discrete task was written on a paper card and then matched with the mission area or areas in which the task needed to be performed. This exercise helped uncover that some tasks spanned more than one mission area and some tasks had to be repeated. At the end of the exercise, all of the discrete tasks were organized into groups and the structure became clear. While the card sorting exercise did not provide the final structure, it did help answer many questions that were useful in tackling the next design phases.

The iterative design then continued as the design team created and analyzed many rounds of mockups of the proposed system, starting with rudimentary drawings eventually progressing to more interactive prototypes. As the design matured, it was handed off to the software development team who coded the functionality for evaluation. During the evaluation phase, several formal desktop usability tests were performed and the findings were included as improvements in the next software build. The evaluation phase ended with the design team and test team using the Marlin user interface to conduct an actual end-to-end mission at sea.

The stakeholder interviews revealed that the design needed to incorporate physical launch and recovery and shipboard cradle systems, as well as software elements that give the operator status from onboard sensors as the AUV performs inspections and creates 3D models of the subsea structure. Via contextual inquiries and observations, it was uncovered that operators assemble a mission plan prior to the mission and then observe the vehicle as it creates models from 3D sonar data on the first mission, followed by change detection missions. The user interface design requirements for the Marlin AUV system thus included the need to account for operators working onboard a ship in an operations and maintenance van, which amounted to designing to account for the crew working in a small office space inside of a shipping container. Stakeholder interviews also uncovered that the maintenance van is hot and dimly lit and the crew sometimes works up to 12-hour shifts while the Marlin is performing subsea inspections. Table 7.2 provides more detail and shows the methods that were employed during each stage.

The UCD design team encountered many design challenges, most notably the consolidation of the existing systems into a single display to account for all phases of vehicle operation. This was due to the underlying architecture that was already in place and could not be changed to meet operator tasking that was discovered in the user research phase. The existing systems were architected and built as pure engineering test beds while the goal of the new system was an operator task-centric focus. However, following the UCD process, and selecting the best methods as the design progressed, the team was ultimately successful in delivering a user interface that met the requirements. Figure 7.5 shows the design progression, starting with an original screen from one of the legacy systems, ending with the end user interface design.

## FUTURE TRENDS

Every year, there are numerous articles published online with predictions for UX trends. Some predictions are realized, while others fade into the background as new technologies emerge. If the past is a good predictor of the future, the future of UX will be tightly coupled with emerging technologies. Gartner's Hype Cycle (Gartner, 2015) is a good resource for understanding emerging technologies and their expected impacts. The Hype Cycle includes things that have already started to be realized, such as 3D printing, gesture controls, and autonomous vehicles. It also includes things that are still in their infancy and considered innovation triggers, such as virtual personal assistants, brain computer interfaces, and smart robots. The implications for user interfaces and UXs of these emerging technologies are profound and the field of UX will need to adjust accordingly. Indeed, the introduction of these technologies may change the way that people expect to interact with existing technologies, and therefore, UX practitioners may need to reevaluate UXs for existing products and systems in light of these new perspectives and expectations.

The future of UX is also likely tied to the future of systems/software development. UX practitioners have been greatly affected by the move from waterfall to agile development, which has forced UCD practitioners to significantly change how we approach problem spaces to fit within the new project structure and schedules. For example, while in agile development methods there is typically not a plan to iterate within a sprint when taking a UCD approach to agile development, UCD practitioners need to communicate to agile developers that it is critical to make time to iterate so that designs can respond to user feedback. If the development world moves toward a different style that varies from agile or waterfall, UX practitioners will need to adjust once again to continue to stay relevant and successful in the future.

## TABLE 7.2
## UCD Applied to the Marlin Project

| Phase | Methods Used | Real World Factors That Influenced Methods |
|---|---|---|
| User research | Review of current system<br>End users and stakeholders<br>–Interviews with the test<br>team<br>–Collaborative working<br>groups<br>–Off-shore test inspection<br>observations | *–Availability of working displays*—The existing GUIs that were used by the test team were available. These existing GUIs, while not user friendly, revealed essential features and functions that needed to be represented in the final design.<br>*–Access to current in-house operators*—Interviewing the test team yielded rich data from which to start the design. They were able to share what they needed most to operate the vehicle, what they would only need to troubleshoot software issues behind the scenes, and the functionality that is critical to ensure the vehicle is operating safely and efficiently.<br>*–Access to current team members*—Collaborative working groups with stakeholders including Business Development and the Leadership Team yielded information about future enhancements that were envisioned to accommodate additional missions and customer data analysis.<br>*–Access to current in-house operators*—Observing the test team operating the vehicle using the existing GUIs uncovered pain points and allowed the human factors team to observe first-hand the most frequently used data. |
| Iterative design | –Collaborative working<br>groups<br>–Information architecture<br>card sorting exercise<br>–Initial storyboarding<br>–Navigational workflow<br>–Mockups and interactive<br>prototypes<br>–Detailed interaction<br>design | *–Access to current in-house operators*—As the design emerged, collaborative working groups with stakeholders and the test team were ongoing. During these sessions, emerging designs were iterated via end-to-end mission scenario walk-throughs, which helped flesh out the design from the information architecture to the most optimal individual widget placement.<br>*–Frequent peer reviews*—After feedback from the collaborative working groups was folded back into the designs, human factors engineers provided feedback from a best practices perspective to the lead designer. |
| Evaluation | End users<br>–Weekly peer reviews/<br>heuristic evaluation<br>–Desktop usability<br>evaluations using a<br>modified version of the<br>RITE (Rapid Iterative<br>Testing and Evaluation)<br>method<br>–Offshore test events | *–Access to current in-house operators*—The peer reviews were conducted by engineers who had knowledge of the emerging design but were not involved in the minutiae of the design, so that they had knowledge to ask questions and make feasible suggestions.<br>*–Tight development timeline*—Due to time constraints, usability tests were conducted on actual production code vs. prototype. |

*(Continued)*

**TABLE 7.2 (CONTINUED)**
**UCD Applied to the Marlin Project**

| Phase | Methods Used | Real World Factors That Influenced Methods |
|---|---|---|
| | | –*Scheduling difficulties*—Scheduling participants with experience operating the vehicle proved difficult because of the small population. For one usability test, much of the participant pool was unavailable due to off-shore inspections in the Gulf. Of those that were recruited, only half of the participants had hands on experience; for another, none had direct experience but all were in the process of completing vehicle operation training. |
| | | –*Tight development timeline*—Technical difficulties led to one participant joining another participant's session and the two participants jointly conducted the usability testing tasks. |
| | | –*Need for realistic mission scenarios*—The order of tasks had to be the same for each session to follow the realistic sequence of events. |
| | | –*Testing and verification*—Offshore test events were conducted where the lead designer and some contributors to the design were given the opportunity to use the new GUI to operate the vehicle. |
| | | –*Tight development timeline*—It was difficult to gauge usability for multiple reasons: (a) many bugs, (b) workflows not fully fleshed out, and (c) some things were not yet implemented. |
| | | –*Tight development timeline*—The development team provided a representative to observe testing and be available for technical assistance, which proved to be valuable while working directly with the production code. |
| Beyond | –Offshore operations with live system<br>–Worked with software team to build the system by analyzing tradeoffs and ensuring design concept is realized | –*Testing and verification*—Offshore test events continued to be conducted after the design was agreed to in order to rectify functionality that did not work as envisioned during the design phases.<br>–*Additions of new functionality*—Follow-on work to add new functionality and enhance existing baseline functionality considered tradeoffs with regards to design and technology decisions as enhancements and additions were iterated through the UCD process. |

# CONCLUSION

This chapter provided a realistic overview of how practitioners apply UCD in their real-world projects, using the Lockheed Martin Marlin UAV as a case study. In addition to demonstrating the outcome of that project, the chapter provided insights into decisions made along the way to ensure a successful final design given all of the

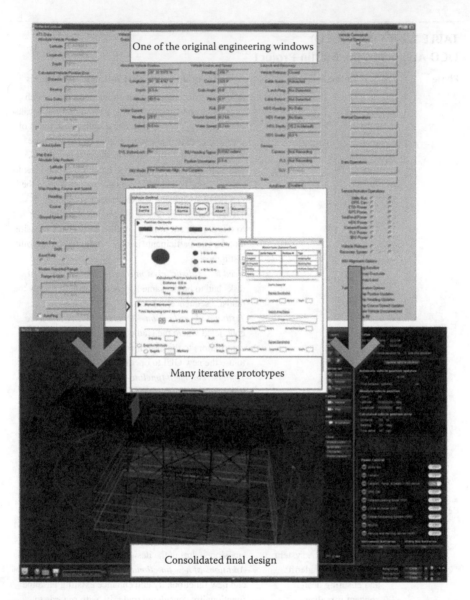

**FIGURE 7.5** Progression of Marlin user interface designs. Copyright 2017 by the Lockheed Martin Corporation.

real-world constraints of the project. It is important for practitioners to understand that the ideal UCD project is very rare and that flexibility and adaptability are key factors in keeping the end user needs and perspectives at the heart of the design process in each phase in whatever manner is realistic. UCD is about focusing on the end user's needs to the extent possible, using whatever resources are realistic and available to help achieve that goal.

## REFERENCES

Casey, S. (1998). *Set phasers on stun and other true tales of design, technology, and human error.* Santa Barbara, CA: Aegean Publishing Company.

Casey, S. (2006). *The atomic chef and other true tales of design, technology, and human error.* Santa Barbara, CA: Aegean Publishing Company.

Courage, C. & Baxter, K. (2005). *Understanding your users: A practical guide to user requirements.* San Francisco: Morgan Kaufmann.

Eberts, R. E. (1994). *User interface design.* Englewood Cliffs, NJ: Prentice Hall.

Gartner. (2015). *Emerging technology hype cycle.* Retrieved on June 1, 2016, from http://www.gartner.com/newsroom/id/3412017

Gould, J. D. & Lewis, C. (1985). Designing for usability: Key principles and what designers think. *Communications of the ACM, 28*(3), 300–311.

Kjær-Hansen, J. (1999). *A business case for human factors investment.* European Organisation for the Safety of Air Navigation. Report HUM.ET1.ST13.4000-REP-02. Retrieved from http://www.idemployee.id.tue.nl/g.w.m.rauterberg/lecturenotes/HF99cost-report.pdf

Lidwell, W., Holden, K., & Butler, J. (2010). *Universal principles of design, revised and updated: 125 ways to enhance usability, influence perception, increase appeal, make better design decisions, and teach through design.* Gloucester, MA: Rockport Publishers.

Martin, B. & Hanington, B. (2012). *Universal methods of design: 100 ways to research complex problems, develop innovative ideas, and design effective solutions.* Gloucester, MA: Rockport Publishers.

Mohammed, N., Munassar, A., & Govardhan, A. (2010). A comparison between five models of software engineering. *International Journal of Computer Science Issues, 7*(5), 94–101.

Norman, D. A. (1986). Cognitive engineering. In D. A. Norman & S. W. Draper (eds.), *User centered system design: New perspectives on human–computer interaction* (pp. 31–61). Hillsdale, NJ: Lawrence Erlbaum.

Palmquist, M. S., Lapham, M. A., Miller, S., Chick, T., & Ozkaya, I. (2013). *Parallel worlds: Agile and waterfall differences and similarities.* Pittsburgh, PA: Carnegie Mellon University Software Engineering Institute.

Spath, D., Hermann, F., Peissner, M., & Sproll, S. (2012). User requirements collection and analysis. In G. Salvendy (Ed.), *Handbook of human factors and ergonomics* (4th ed., pp. 1313–1322). Hoboken, NJ: John Wiley & Sons.

Stone, D., Jarrett, C., Woodroffe, M., & Minocha, S. (2005). *User interface design and evaluation.* Los Altos, CA: Morgan Kaufmann.

Wilson, C. E. & Rosenbaum, S. (2005). Categories of return on investment and their practical implications. In R. G. Bias & D. J. Mayhew (Eds.), *Cost-justifying usability: An update for the Internet age* (pp. 215–263). San Francisco: Morgan Kaufman Publishers.

## KEY TERMS

**iterative design:** a series of increasingly more detailed and complete attempts at a user interface design (i.e., low-fidelity sketches to high fidelity, interactive prototypes), until a design emerges that meets some threshold of acceptance specified by your target customer and your own team.

**prototyping:** tangible examples of design concepts created to gather feedback from team members, stakeholders, end users, or anyone who may have valuable insights about an interactive technology, range from low-fidelity paper-and-pencil sketches to high-fidelity, fully interactive software interfaces.

**usability testing:** evaluation technique in which the usability of a system is measured by evaluating representative end users while they interact with a user interface in order to uncover the strengths and weaknesses of the design.

**user-centered design:** an approach to the design of interactive technologies that focuses on incorporating the anticipated end users' needs into the user interface at each stage of development to ensure the tasks and workflows in the final system meet usability goals.

**user experience design:** an extension of user-centered design that places more emphasis on the overall experience surrounding the human–computer interface, the context in which the interaction will occur, the users' affective responses, etc.

**user interface design:** user interface design focuses on anticipating tasks and workflows that users will perform and ensuring that the interface has elements that are easy to access, understand, and use to facilitate those actions.

**user requirements:** detailed software specifications that identify the functionality that a user needs and/or expects a user interface to perform to complete some task(s).

# 8 Workspace Design

*Caroline Joseph*

## CONTENTS

## INTRODUCTION

Workers do not operate independently of their environment but interact with it as part of their daily work. Workers are therefore influenced by workspace design. The way the environment is designed and maintained (i.e., workspace design) will dictate how productive workers are but also whether they are at risk of getting injured or becoming ill. For example, if workers need to always reach far away to complete their work it could slow them down and increase their risk of getting injured over time. Having a workspace designed for workers will allow them to complete their tasks safely, effectively, and efficiently. It allows them to stay motivated and, in the long run, allows them to stay at work up to retirement. Having a productive workforce is important, but if the well-being of workers (health and safety) is compromised, production is likely to decrease and profits will most likely be affected as well.

The financial impact of occupational injuries and illnesses was estimated at approximately $250 billion (both indirect and direct costs) in 2007 in the United States alone (Leigh, 2011). We often think the only costs of injuries are the costs for health care treatments. However, the financial ramifications are often much larger. Lower productivity due to a decreased attendance, increased training costs (if a worker is absent for a prolonged period), and increases in insurance contributions are a few of the factors that may influence the total cost of an injury. To reduce the financial burden of occupational injuries and illnesses, workspaces need to be designed, evaluated, or improved to ensure that no harmful condition exists.

In this chapter, we will describe factors and conditions to consider when designing workspaces. We will highlight tools and methods that can be used to quantify the exposure to risk factors for occupational injuries and illnesses and therefore inform where changes should be done. We will present examples of applications of the tools and methods along with future trends in ergonomics and occupational safety.

## FUNDAMENTALS

In this section, we will cover the relation between the individual's capacity and demands imposed by the environment and work tasks. We will explore how the interaction between capacity and demand can lead to injury, the main contributing factors to occupational injuries and illnesses, and which contributing factors can be quantified. Finally, we will identify elements to pay attention to when designing or improving a workspace that will facilitate workers returning to work more rapidly after an injury and staying at work injury-free for their full career.

When work characteristics (demands) exceed what the body can handle, an injury can occur. This concept is often referred to as a mismatch between the capacity of a worker and the demand of a work task. A worker's capacity is what his or her body can handle and produce—that is, its characteristics, including limitations, and capabilities or abilities. Capacity includes the worker's anthropometry, strength, range of motion, and endurance. No two workers have the same capacity. To know each worker's individual capacity, health practitioners can perform functional capacity evaluations. As ergonomists and safety professionals rarely have access to a worker's exact capacity, they need to rely on general population estimates. Databases (Pheasant & Haslegrave, 2006) are available that include distributions, often stratified by age and gender, for anthropometric variables such as height, limb length, grip strength, pinch strength, and heart rate.

Conversely, work task demands are the requirements imposed by the work task on the person performing it. These demands include work characteristics and the minimum capacity required to perform the task. Work task demands commonly include pace, duration, force application, and working posture requirements. Work task demands are evaluated using physical demands analysis and other methods that will be highlighted in the following section.

Depending on the nature of the mismatch between capacity and demand, acute injury, cumulative injury, or illness can occur. *Acute injuries* occur when the work demands exceed worker capacity or if there is an acute event that leads to an immediate temporary or permanent modification of the body. Examples of acute injuries are fractures, lacerations or burns, sprains, and strains. In these cases, the body does not have adequate capability to resist the sudden excessive demands. *Cumulative injuries* occur when the demand is close to the worker capacity but there is an extended exposure to the demand that leads to a temporary or permanent modification of the body. Examples of cumulative injuries are tendonitis, bursitis, back pain, carpal tunnel syndrome, and hearing loss. Such cases usually occur when the body is unable to recover between exposures, leading to the injury over time. Musculoskeletal disorders are a broad category of cumulative injuries to the bones, soft tissues (tendon, ligament, and bursa), and muscles. *Occupational illnesses*, similarly to cumulative injuries, occur when the exposure at work leads to chronic diseases over time. Examples of occupational illnesses are lung diseases, cancer, and dermatitis.

The cause of occupational injuries and illnesses is usually multifactorial. The contributing factors or risk factors can be grouped in four categories: individual, physical, psychosocial, and environmental. *Individual* factors are personal characteristics of the workforce and their capacity. Some examples are the health status of workers, preexisting conditions (illnesses and injuries), physical capacity (e.g., strength and endurance), height, weight, gender, habits (e.g., smoking), and age. Other than in the case of personal habits and to some extent physical capacity, individual factors cannot be changed. *Physical* factors are sometimes referred to as physical exposure and relate to the physical design of the workspace, how work is performed, and the equipment used. These factors are also the main component of the physical demand of the work. Some examples are forces that need to be exerted or postures that workers need to adopt to complete a task, the presence of vibration, and repetition of specific movements. *Psychosocial* factors are related to social interactions and how the work environment can affect individuals psychologically. Some examples are stress level, the type of company management, level of control on work, and interaction between workers. Finally, *environmental* factors relate to the work environment in general and cover factors not included in the previous three categories. These factors are another component of the physical demand of the work. Some examples are workspace illumination, noise levels, chemicals exposure, and air quality.

While most of the risk factors for occupational injuries and illnesses can be quantified, not all of them have known threshold values that will lead to temporary or permanent injury or illness. Most threshold values have been established when single risk factors are the cause for an injury or illness. For chemical exposure, threshold values or level of risk for different types of injury or illness are available in a product's Safety Data Sheet provided by product manufacturers. For other risk factors such as noise levels, vibration, lighting, or oxygen levels in confined spaces, thresholds and/or recommended values are available in the literature but most importantly in local Occupational Health and Safety Regulations, Acts, and Laws. For musculoskeletal disorders, although individual risk factors can lead to injuries, in most cases, it is the influence of multiple risk factors acting simultaneously that lead to an injury. For example, there is evidence that excessive force, repetitive work, and extended work in nonneutral postures lead to musculoskeletal disorders (National Institute for Occupational Safety and Health [NIOSH], 1997). Further, combining risk factors such as high force application with highly repetitive work will amplify the risk of injury as compared to each factor individually (Silverstein, Fine, & Armstrong, 1987). Consequently, the presence of multiple risk factors can drastically increase the likelihood of injuries and illnesses as compared to the presence of a single risk factor. Kodak's *Ergonomics Design for People at Work* (Chengalur, Rodgers, & Bernard, 2004) and *Fitting the Task to the Human* (Kroemer & Grandjean, 1997) are excellent resources for practitioners on specific design guidance, recommendations, and threshold values that could be used in the field.

The best way to keep workers injury-free, healthy, productive, and happy is to have a workspace that is designed for the workers and can accommodate the workforce over time. These workspaces are designed to accommodate for age-related changes that may include decreased capacity (reduction in strength, range of motion, etc.). Often, practitioners will use anthropometric databases and consider the capacity of small females (5th female percentile) and large males (95th male percentile) in their design to try to accommodate the majority of the population. For example, using the anthropometric information of large males can help with identifying what should be the minimum ceiling height of a work environment with limited space, while using the anthropometric information of small females can help define what should be the maximum strength or force requirement for a task. Consequently, adopting such an approach can result in a workspace that can accommodate changes in capacity over time.

Although attempts are made to design workspaces and work tasks within the worker's capacity, occupational injuries and illnesses are still common. Injuries not only lead to loss of income but they could also have an influence on employee morale. One solution to this problem is to have a system in place to allow for a quick return to work. In many cases, health practitioners will allow a prompt return to work but with modified duties—that is, duties or tasks within the injured worker's abilities are assigned to him or her since the demand of regular work is too high at the moment. In reassigning an injured worker to a new task, employers need to consider the worker's usual work. For example, if a worker usually performs physically active work, he or she can be temporarily assigned into an easier but still physically active work instead of desk work. To ensure that modified work assignments are effective, employers must know the general physical demand of most, if not all, work tasks. This makes it possible to find comparable work tasks, based on task descriptions, which can accommodate the worker's temporary capacity.

## METHODS

Methods to evaluate work and workspaces are often derived from known risk factors for musculoskeletal and other disorders. A number of methods are available to evaluate workspaces and to prevent musculoskeletal disorders. It is beyond the scope of this chapter to elaborate on each method; instead, a brief overview of commonly used methods to assess workspaces is presented. No single tool or method covers all risk factors and most ergonomists use multiple methods simultaneously (Dempsey, McGorry, & Maynard, 2005). Methods are selected based on the information they will provide while acknowledging their limitations. The information and conclusions gathered through different tools and methods can then help to inform changes to the workspace to help reduce the risk of injury to specific body areas.

Subjective evaluation of discomfort and symptoms associated with work related-musculoskeletal disorders can provide a wealth of information. This can be used to prioritize work tasks for evaluation, to help with the selection of evaluation techniques, to track improvement or deterioration over time, and to document the influence of any intervention. A simple approach is to use a body part discomfort map, where workers subjectively rate the intensity of discomfort for specific body parts

(Corlett & Bishop, 1976). Discomfort is usually quantified for each body on a 5- or 7-point scale from "no discomfort" to "extreme discomfort." The Standardized Nordic Questionnaire can also be used to evaluate symptoms of musculoskeletal disorders in the previous 7 days and the previous 12 months for the neck, upper back, lower back, upper arms, and lower arms (Kuorinka et al., 1987).

The most important part of any workspace evaluation is to gain a thorough understanding of the work itself. Different methods can be used to understand the work, but conducting a task analysis that hierarchically breaks down the work into its elements, both physical and cognitive, is an excellent first step. This is often done by referring to standard operating procedures and by talking to, and observing, workers perform the work. Referring to standard operating procedures helps to clarify how the work should progress. Interacting with workers helps to identify how work is carried out in practice and identify variations or deviations from standard operating procedures.

Observation, both real-time and simulated real time (using collected video), can be used by experienced observers to evaluate task time, record working posture, and identify other risk factors. Recording video to review and analyze later can be beneficial. However, one caveat is that analysis can be time consuming and if video is not collected appropriately (e.g., bad observation angles, shaky video, blurred video), analyses can be a challenge (Heberger, Nasarwanji, Paquet, Pollard, & Dempsey, 2012).

The gold standard is to record data (e.g., posture, motion, muscle activity) directly and continuously from the workers. Direct measures provide accurate measurements of the influence of the work demands on workers. Direct measurements can be a challenge for practitioners as they may not have the required tools and resources to do so. Some shortcomings of direct measurements are that the measuring equipment is often sensitive to the environmental condition and it can interfere with the way workers perform their work. The trade-off between the benefits versus time and effort versus shortcomings of using direct measurements should be considered prior to their use.

## GENERAL WORK EVALUATION METHODS

Prior to beginning work, it is preferable to identify and keep a record of the physical work requirements to ensure the individual is capable of doing the work, with or without accommodations, and can do so without getting injured. One method commonly used to evaluate work requirements is to perform a physical demands analysis (Workplace Safety & Prevention Services [WSPS], 2011). A physical demands analysis describes primary work tasks, lists tools and equipment that would be used, quantifies the frequency of use and type of postures for all body parts, quantifies force (e.g., lifting, carrying, pushing, pulling, gripping) requirements along with their frequency, lists environmental factors that may be encountered (such as noise and vibration), and lists other demands such as sensory demands, psychosocial factors, and mobility requirements that may be encountered as part of the work. This, in conjunction with a functional capacity evaluation (Chen & Joseph, 2007), can ensure the work is designed appropriately for the individual.

A number of observational tools are available to evaluate risk factors for musculoskeletal disorders once at work. Most of these tools evaluate a combination of risk factors including posture, force, and repetitiveness, along with other specific risk factors such as work pace, duration, and type of grip. These tools range from quick assessments such as the Quick Exposure Checklist (QEC) (Li & Buckle, 2005), Plan för Identifiering av Belastningsfaktorer (Plan for Identification of Load Factors) (PLIBEL) (Kemmlert, 1995), and Occupational Repetitive Actions (OCRA) (Occhipinti, 1998) to more involved methods that require a thorough evaluation of work and working posture such as the Ovako Working Posture Analysis (OWAS) (Kivi & Mattila, 1991; Mattila, Karwowski, & Vilkki, 1993) and Posture Activity Task and Material Handling (PATH) (Buchholz, Paquet, Punnett, Lee, & Moir, 1996). In addition, some tools focus on specific body areas, such as the Rapid Upper Limb Assessment (RULA) (McAtamney & Corlett, 1993), while others are more inclusive such as Rapid Entire Body Assessment (REBA) (Hignett & McAtamney, 2000). Table 8.1 provides a list of a few tools along with risk factors they consider and data collection strategies. These tools are often used by researchers and practitioners in the field as they are simple to use, provide a final score or a recommendation on whether the work is a risk, and can often shed light on specific parts of the work that need improvement. Tools that provide a score or specific recommendations are beneficial to practitioners as they can be used to justify expenditure to improve workspaces and can be used to track changes over time. However, each tool has its strengths and weaknesses. Hence, they cannot be used in all settings or for all tasks and practitioners who use the tools should be adequately trained to ensure valid results.

## TABLE 8.1
## Description of Observational Tools Including Risk Factors Considered and Data Collection Strategies

| Tools | Risk Factors Considered | Body Areas Considered | Data Collection Strategy |
|---|---|---|---|
| Quick exposure check (QEC) | Posture Force Duration Frequency Movements | Back Shoulders and arms Wrists and hands Neck | "Worst case" of the task |
| Plan för Identifiering av Belastningsfaktorer (Plan for Identification of Load Factors) (PLIBEL) | Posture Force Frequency Movements | Neck/shoulder and upper back Shoulders, forearms, and hands Feet Knees and hips Lower back | Selection by general knowledge of work and observation |
| Occupational repetitive actions (OCRA) | Posture Force Duration Frequency Vibration | Hands Wrists Forearms Elbows Shoulders | Assessment of repetitive action included in profile of work |

*(Continued)*

**TABLE 8.1 (CONTINUED)**
**Description of Observational Tools Including Risk Factors Considered and Data Collection Strategies**

| Tools | Risk Factors Considered | Body Areas Considered | Data Collection Strategy |
|---|---|---|---|
| Ovako working posture analysis (OWAS) | Posture Force | Back Arms Legs | Time sampling |
| Posture activity task and material handling (PATH) | Posture Force Work activity | Back Neck Legs Arms | Time sampling |
| Rapid upper-limb assessment (RULA) | Posture Force Static action | Wrists Forearms Elbows Shoulders Neck Trunk | No detailed rules |
| Rapid entire body assessment (REBA) | Posture Force | Wrists Forearms Elbows Shoulders Neck Trunk Back Legs Knees | Most common/ prolonged/loaded/ postures |

*Source*: Adapted from Takala, E. P. et al., *Scandinavian Journal of Work, Environment & Health*, 36(1), 3–24, 2010; Imbeau, D. et al., Troubles musculo-squelettiques: évaluation et conception du travail. In *Manuel d'hygiène du travail: Du diagnostic à la maîtrise des facteurs de risque* (pp. 321–362), Roberge, B. et al. (eds.), Mont-Royal: Modulo-Griffon (in French), 2004; Joseph, C., *Impact des transformations réalisées dans le cadre d'un projet d'amélioration continue sur l'exposition aux TMS (troubles musculo-squelettiques) aux membres supérieurs des scieurs* (unpublished master's thesis). Montréal, Canada: École Polytechnique de Montréal (in French), 2005.

## SPECIFIC EVALUATION METHODS

When performing specific evaluations of work, two primary areas of interest are posture and motion and physical exertion and force.

### Posture and Motion

Static posture over an extended period of time or repetitive use of nonneutral posture can pose a risk for musculoskeletal disorders. The simplest method to evaluate posture is to observe workers and identify nonneutral postures. A few of the rapid assessment methods described earlier could also be used. When postures are observed, they are often coded above or below a threshold, say trunk flexion greater than 20°, or into categories, such as shoulder flexion between 0°–30°, 30°–60°, 60°–90°, or greater than 90°. When coding posture, it is essential to have trained observers and preferably more than one to ensure reliable and valid postural coding. To improve the accuracy of postural measurements, a goniometer (similar to a protractor) can be used. To directly measure posture, workers can be instrumented with electrogoniometers that provide a continuous measurement of working posture. Electrogoniometers are devices with small strain gauges that are mounted across a body joint to measure its movement. As posture is recorded continuously, electrogoniometers can also provide kinematic information of motion across the joint. The Lumbar Motion Monitor is an electrogoniometer designed to measure low back flexion, extension, lateral bending and rotation, and has been successfully used in the field to evaluate work, especially manual material handling (Marras et al., 1993).

In addition to measuring joint posture, it is often of interest to evaluate overall body motion to quantify the repetitiveness of work or identify kinematic variables such as acceleration and velocity. Motion tracking devices can track an individual's posture while he or she does work. Most motion tracking systems either use line of sight to track active or passive markers attached to the person or small inertial measurement units attached to the body that calculate positions relative to one another.

## Physical Exertion and Force

Physical exertion or the amount of energy used while performing work can provide an indication of the physical stress placed on the body. Measuring the heart rate is used not only by ergonomists but also exercise physiologists to monitor physical exertion. Although a lot of work tasks evaluated are not extremely strenuous, an elevated heart rate can indicate which specific parts of the work are challenging, and it can be used to study the influence of different work techniques and environment on workers. If it is not possible to measure the heart rate, validated subjective ratings of exertion can be used. For example, the Borg Rated Perceived Exertion is a scale that has been shown to be correlated to the heart rate (Borg, 1982, 1990). It is a continuous scale used to rate exertion from a minimum value of 6 (*no exertion at all*) through 13 (*somewhat hard*) through a maximum value of 20 (*maximal exertion*). As an alternative, the Borg CR-10 scale can be used to measure exertion (Borg 1982, 1990). It is a continuous scale used to rate exertion from a minimum value of 0 (*nothing at all*) through 5 (*strong (heavy)*) through 10 (*extremely strong (maximal)*) through a maximum value anywhere above 10 (*absolutely maximum (heaviest possible)*).

Measuring the force applied can be used to estimate biomechanical loads on the body. The actual force applied can be measured by instrumenting equipment with strain gauges and having workers perform or simulate the task of interest. Dynamometers can be used to measure pushing, pulling, lifting, lowering, grasping, and pinching forces directly by attaching them to the equipment. If a thorough understanding of how the body interacts with the work environment to generate force is required, electromyography can be used to measure the muscle electrical activity using small electrodes. Although muscle activity is not a direct measure of muscle force, it can provide an estimate of how much that individual muscle is being activated or used. Entire books are available on explaining electromyography (Basmajian, 1962; Criswell, 2010) due to the complexity associated with instrumentation, data collection, and analysis. However, measuring muscle activity is commonly adopted to evaluate work tasks that predominantly use the upper extremities and/or the back. When designing new tasks, it can be a challenge as there is no way to estimate required force and loads. Psychophysical tables for acceptable lifting, lowering, pulling, and pushing loads for various durations can be used as a starting point when designing work (Ciriello & Snook, 1983; Ciriello, Snook, & Hughes, 1993; Liberty Mutual, 2007; Snook, Vaillancourt, Ciriello, & Webster, 1995).

An area of special interest in work evaluation is manual material handling as it has been a large contributor to lower back injuries. Any manual material handling task that involves lifting or lowering with a weight of 50 lbs or more (Occupational Health and Safety Administration [OSHA], 2015) should be avoided, and lifting should be carried out in the power zone which is between the waist and the shoulders. In addition, trunk rotation (twisting) or lateral bending should be avoided when lifting as it increases the risk for injury. The NIOSH lifting equation can be used to identify if the lifting task poses a risk for lower back injuries (Waters, Putz-Anderson, Garg, & Fine, 1993). The NIOSH lifting equation considers the height from which the object is lifted, the distance lifted, the distance from the body, the frequency of lifting, if the lift requires trunk rotation or bending, and the type of coupling. Software alternatives to evaluate material handling tasks are 3D SSPP (Chaffin, Andersson, & Martin, 1999) and 4D Watbak (Neumann, Wells, & Norman, 1999).

## APPLICATION

In practice, not all organizations have the same view on workspace safety and ergonomics. Some organizations are proactive and will take measures to prevent injuries while other organizations are reactive and will wait for an injury to occur. Proactive organizations will use some of the methods presented in the preceding section in conjunction with ergonomics principles to design tasks and workspaces that will not exceed workers' capacities. Reactive organizations will also use some of the methods presented above to help modify problematic tasks or workspaces, or to help inform their return-to-work program. Ideally, it would be preferable to begin with perfectly designed workspaces, but with fast-paced changing industries, this is not always possible. In this section, we will present two examples of application of some of the tools and methods described previously: one in food manufacturing and one in construction.

### EXAMPLE IN FOOD MANUFACTURING

This first example shows how some of the methods described in the previous section were used at a fish processing manufacturing plant. An audit from the local OSHA board determined that changes were required at a workspace where large frozen fish blocks were cut into smaller blocks fed into an automatic multisaw for portioning to reduce the likelihood of developing musculoskeletal disorders. This work was separated into four tasks, each performed by a different worker:

1. Unpacking fish blocks from a box (three blocks per box),
2. Removing a wax cardboard protecting the individual blocks,

3. Cutting a first slice of the block using a band saw, and
4. Cutting the rest of the block into smaller blocks using a band saw and feed-
   ing them into the multisaw.

To gain a thorough understanding of the problem at hand, three tools (presented in the Methods section) were used for the baseline assessment: QEC, OCRA,
and 4D Watbak. The baseline assessment indicated that the main problems were the
weight of the frozen fish blocks manipulated (two or three blocks were manipulated
at a time) and the height at which the blocks were stacked on buffer tables positioned
beside each band saw. Consequently, the main objective of the first round of modifications was then to reduce the maximum weight manipulated/lifted and to reduce the
height for the lifts. The solution was to adjust the height of the buffer tables, to place
shorter stoppers to ensure that workers could not stack blocks as high (two blocks high
instead of four), and to implement a policy where workers were required to only lift
one block at a time instead of multiple ones. Further assessment with the same tools
showed that these modifications directly addressed the issues raised in the OSHA
audit and reduced the likelihood of injury as compared to the baseline assessment.

Although the organization had remediated the original issues identified, it decided
to be proactive and make a second round of modifications. As part of the second
round of modifications, the organization decided to take a closer look at the entire
work process, in addition to individual work tasks. By looking at the big picture, it
was able to eliminate one sawing operation on the first band saw. In the meantime,
the organization also started receiving the individual fish blocks stacked on pallets
instead of boxes from the supplier. This removed the need to remove blocks from
the boxes. Additional assessments showed that both rounds of changes dramatically
reduced the level of exposure to risk factors for musculoskeletal disorders.

## EXAMPLE IN CONSTRUCTION

Unlike the previous example, in a traditional manufacturing setting, the approach
to workspace design in construction needs to be different. This is because construction projects have a fixed duration (usually short to moderate length), projects have a
rotating workforce (usually local), the workspace is constantly evolving throughout
the project, and no project is exactly the same as another. For these reasons, taking
the time to evaluate all the small details of a work task might not be as relevant
since the exact same task in the exact same conditions might not recur. That said, it
is still important to have an understanding of the work and risk factors present, to
not only prevent injury but also to help assign or reassign workers appropriately. As
construction projects usually have tight schedules, losing workers temporarily due
to an injury can be challenging, especially if they are skilled workers. One way to
dampen this effect is to have the worker promptly returning to work in a task that
he or she can perform. Establishing which tasks could be performed and knowing
the demands associated with the work tasks can help with worker reintegration. To
help inform its return-to-work program, a large general contractor involved with the
construction of wind turbines decided to use a physical demand analysis to evaluate
some of its work operations.

Once a wind turbine is fully assembled and is mechanically complete, it usually needs to be cleaned before custody is transferred to the customer. It was therefore decided to evaluate the physical demand of the different subtasks involved with cleaning a turbine. The evaluations were done on two models of turbine from a single turbine manufacturer since both models were present on the project site. For both turbine models, most tasks were the same, but when differences in the turbine design influenced the work, separate forms were used. Cleaning the turbine was separated into seven tasks based on the nature of the work and the area where the work was performed (Figure 8.1):

- Cleaning the basement (basement design was different between the two turbine models),
- Cleaning large electrical cabinets (only one of the two turbine models),
- Cleaning the decks,
- Cleaning the walls,
- Painting walls where original paint was chipped or damaged,
- Cleaning the nacelle and paint touch-ups, and
- Cleaning the hub.

Of all tasks identified, it was found that the least physically demanding task was to clean the basement of the turbine as it did not require workers to climb up a ladder to the top of the turbine. To accomplish this task, workers either had to open some hatches and crawl into a 3-foot basement or to climb down a 10-foot ladder to vacuum and clean the basement, depending on the turbine model. All other tasks had about the same level of force or strength demand. The main difference in physical demand between tasks was due to the posture the workers had to place themselves into to accomplish the work. Table 8.2 highlights which body areas were mostly engaged by the work. In this table, high back demand relates to constant back movement, high

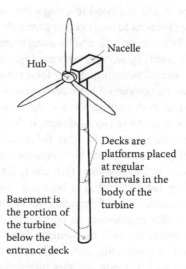

**FIGURE 8.1** Sections of a wind turbine.

**TABLE 8.2**
**Summary of High Postural Demand Tasks while Cleaning a Wind Turbine**

| Task | Back | Shoulder | Arms | Neck | Legs |
|---|---|---|---|---|---|
| Cleaning the basement | | | | | |
| Cleaning large electrical cabinets | | High | | | High |
| Cleaning the decks | | High | | | High |
| Cleaning the walls | High | High | | | |
| Painting walls | | High | | | High |
| Cleaning the nacelle and paint touch-ups | | | | | |
| Cleaning the hub | High | | | | |

shoulder demand relates to frequent or constant reach above shoulder level, and high legs demand relates to frequently or constantly crouching, squatting, or kneeling.

Knowing which body areas are mainly involved for each task, it becomes possible to identify which tasks an injured worker could potentially be temporary reassigned to. For example, if a worker who usually works up in the tower become injured and cannot climb, he or she could be reassigned to clean the basement. Workers who have sensitive shoulders or had shoulder injury before should probably be assigned some of the tasks that are less demanding on the shoulder (e.g., cleaning the basement, the nacelle, or the hub). Similarly, a worker with a sensitive back (or that had back injury before) should probably be assigned to tasks that are less demanding on the back, such as cleaning the basement, cleaning the electrical cabinets, cleaning the decks, painting the walls, or cleaning the nacelle.

## FUTURE TRENDS

With the aging demographic and the need to design the work to be more inclusive, designing work and the workspaces to support employees throughout their entire work life has become essential. This can be achieved by using human factors and ergonomics principles and methods. In addition, workers who are injured and would like to return to work are often not considered. Accommodating for a return to work can benefit not only the individual but also the company. One way to do so is through a comprehensive understanding of the physical work requirements so that work can be designed to support these individuals. Concepts like universal design, inclusive design, or designing for diverse user populations are mainly used in other fields such as architecture, but now these approaches are finding their way into environment and work design as seen in books such as *"Extra-Ordinary" Ergonomics* (Kroemer, 2006). These concepts entail making sure that everything that is designed or built (e.g., product, building, and environment) could be used by people with different abilities (young or old, disabled or not).

To help proactively consider ergonomics as part of the design process, computer-aided design software is available, such as Siemens Jack (Siemens PLM Software, 2016), and commonly used by large companies such as Ford. The Siemens Jack software simulates individuals with varying abilities interacting with a prototype work environment to identify if the individuals have the necessary reach and strength capabilities to complete the task. Although the software is expensive, its benefits are significant as it ensures not only that the designs are adequate but also that workers are less likely to get injured. Smaller companies that do not have the financial resources for such software can mock-up or simulate tasks prior to design and use a human-centered design approach and participatory ergonomics to help eliminate human factors and ergonomics issues during the design phase. Human-centered design approach and participatory ergonomics include workers from the workspace designed or evaluated in the evaluation and modification process. Such approach could then lead to increased workers' compliance with the design and modifications to their workspace.

As the way we think about work, the environment, and its design changes, so do the tools available to measure and evaluate undergoing change. Recent research has tried to use off-the-shelf systems such as the Microsoft Kinect to track posture without any markers on the body (Clark et al., 2012; Dutta, 2012). In addition, the capabilities of smart wearable devices, such as fitness trackers (e.g., Fitbit, Garmin, etc.), are being investigated in novel settings such as to monitor older adults (Lauritzen, Muñoz, Sevillano, & Civit, 2013). Although the accuracy of these new systems is still lacking, research is underway to provide practitioners with tools that are easily available and more cost-effective.

Finally, the omnipresence of smartphones and tablets in the work environment allows practitioners to use them to help improve workspace design. A number of apps designed by research centers, private companies, or individuals are now available to the public at no cost or minimal payment. The range of ergonomic apps available varies in function from those that help workers understand how to use work equipment or safety equipment to detailed ergonomics evaluation. Examples of apps that help workers use specific work equipment or safety equipment are (1) the Ladder Safety app by NIOSH that assists with the safe positioning of the ladder and guidelines for safe use of extension ladders (NIOSH, 2016) and (2) the 3M Respiratory Protection Resource Guide that can be used as a quick reference for the type of respiratory protection to use when chemicals are present in the work environment (3M, 2016). An example of an office ergonomics app is Ergonomics by Stand Up Apps (Stand up, 2016). This app was designed to provide equipment setup advices, stretching exercises, and break reminders. Examples of detailed ergonomics evaluation apps are the 3D SSPP app by the University of Michigan, which is a simplified version of the back biomechanics software described earlier (available on iTunes and Google Play) and the SBN Ergonomics app by Simple But Needed Inc., which covers ergonomics evaluations, site inspection, and lockout tagout (Simple But Needed, 2016).

## CONCLUSION

Workspaces can and should be designed from a human-centric approach to help prevent injury and support a healthy and productive workforce. We introduced the concept of work capacity and demand to highlight how workspace design and workers are interrelated. We also identified factors that should be considered in the design and evaluation of workspaces, including individual, physical, psychosocial, and environmental factors. The examples in the Application section, along with the evaluation methods introduced, can be used as a starting point by practitioners to help proactively design their work environments. Finally, future trends in workspace design have been introduced to encourage practitioners to start or continue designing work for employees' entire working life and support return-to-work programs.

## ACKNOWLEDGMENT

I want to acknowledge the tremendous help of Mahiyar Narsarwanji. His help and input reminded me that a book for practitioners still needs to feature a certain amount of theoretical bases.

## REFERENCES

Basmajian, J. V. (1962). Muscles alive. Their functions revealed by electromyography. *Academic Medicine*, *37*(8), 802.

Borg, G. (1990). Psychophysical scaling with applications in physical work and the perception of exertion. *Scandinavian Journal of Work, Environment & Health*, *16*(Suppl. 1), 55–58.

Borg, G. A. (1982). Psychophysical bases of perceived exertion. *Medicine & Science in Sports & Exercise*, *14*(5), 377–381.

Buchholz, B., Paquet, V., Punnett, L., Lee, D., & Moir, S. (1996). PATH: A work sampling-based approach to ergonomic job analysis for construction and other non-repetitive work. *Applied Ergonomics*, *27*(3), 177–187.

Chaffin, D. B., Andersson, G. B. J., & Martin, B. J. (1999). *Occupational biomechanics* (3rd ed.). New York: John Wiley & Sons.

Chen, M. D., & Joseph J. (2007). Functional capacity evaluation and disability. *The Iowa Orthopaedic Journal*, *27*, 121–127.

Chengalur, S. N., Rodgers, S., & Bernard, T. (2004). *Kodak's ergonomics design for people at work*. Hoboken, NJ: John Wiley & Sons Inc.

Ciriello, V. M., & Snook, S. H. (1983). A study of size, distance, height, and frequency effects on manual handling tasks. *Human Factors: The Journal of the Human Factors and Ergonomics Society*, *25*(5), 473–483.

Ciriello, V. M., Snook, S. H., & Hughes, G. J. (1993). Further studies of psychophysically determined maximum acceptable weights and forces. *Human Factors: The Journal of the Human Factors and Ergonomics Society*, *35*(1), 175–186.

Clark, R. A., Pua, Y. H., Fortin, K., Ritchie, C., Webster, K. E., Denehy, L., & Bryant, A. L. (2012). Validity of the Microsoft Kinect for assessment of postural control. *Gait & Posture*, *36*(3), 372–377.

Corlett, E. N., & Bishop, R. P. (1976). A technique for assessing postural discomfort. *Ergonomics*, *19*(2), 175–182.

Criswell, E. (2010). *Cram's introduction to surface electromyography*. Boston, Massachusetts: Jones & Bartlett Publishers.

Dempsey, P. G., McGorry, R. W., & Maynard, W. S. (2005). A survey of tools and methods used by certified professional ergonomists. *Applied Ergonomics*, *36*(4), 489–503.

Dutta, T. (2012). Evaluation of the Kinect™ sensor for 3-D kinematic measurement in the workplace. *Applied Ergonomics*, *43*(4), 645–649.

Heberger, J. R., Nasarwanji, M. F., Paquet, V., Pollard, J. P., & Dempsey, P. G. (2012, September). Inter-rater reliability of video-based ergonomic job analysis for maintenance work in mineral processing and coal preparation plants. In *Proceedings of the Human Factors and Ergonomics Society Annual Meeting* (Vol. 56, No. 1, pp. 2368–2372). Santa Monica, California: SAGE Publications.

Hignett, S., & McAtamney, L. (2000). Rapid entire body assessment (REBA). *Applied Ergonomics*, *31*(2), 201–205.

Imbeau, D., Nastasia, I., & Farbos, B. (2004). Troubles musculo-squelettiques: Évaluation et conception du travail. In *Manuel d'hygiène du travail: Du diagnostic à la maîtrise des facteurs de risque* (pp. 321–362), Roberge, B., Deadman, J. É., Legris, M., Ménard, L., & Baril, M. (eds.), Mont-Royal: Modulo-Griffon (in French).

Joseph, C. (2005). *Impact des transformations réalisées dans le cadre d'un projet d'amélioration continue sur l'exposition aux TMS (troubles musculo-squelettiques) aux membres supérieurs des scieurs.* (unpublished master's thesis). Montréal, Canada: École Polytechnique de Montréal (in French).

Kemmlert, K. (1995). A method assigned for the identification of ergonomic hazards—PLIBEL. *Applied Ergonomics*, *26*(3), 199–211.

Kivi, P., & Mattila, M. (1991). Analysis and improvement of work postures in the building industry: Application of the computerised OWAS method. *Applied Ergonomics*, *22*(1), 43–48.

Kroemer, K. H. (2006). *"Extra-Ordinary" Ergonomics.* Boca Raton, FL: Taylor & Francis Group.

Kroemer, K. H., & Grandjean, E. (1997). *Fitting the task to the human, Fifth edition—A textbook of occupational ergonomics.* Philadelphia, PA: Taylor & Francis.

Kuorinka, I., Jonsson, B., Kilbom, A., Vinterberg, H., Biering-Sørensen, F., Andersson, G., & Jørgensen, K. (1987). Standardised Nordic questionnaires for the analysis of musculo-skeletal symptoms. *Applied Ergonomics*, *18*(3), 233–237.

Lauritzen, J., Muñoz, A., Sevillano, J. L., & Civit, A. (2013). The usefulness of activity trackers in elderly with reduced mobility: A case study. *Studies in Health Technology and Informatics, 192*, 759–762.

Leigh, J. P. (December 2011). Economic burden of occupational injury and illness in the United States. *Milbank Quarterly, 89*(4), 728.

Li, G., & Buckle, P. (2005). Quick exposure checklist (QEC) for the assessment of workplace risks for work-related musculoskeletal disorders (WMSDs). In *Handbook of human factors and ergonomics methods* (pp. 6-1–6-10), Stanton, N., Hedge, A., Brookhuis, K., Salas, E., & Hendrick, H. (eds.), Boca Raton, FL: CRC Press.

Liberty Mutual. (2012). Tables for evaluating lifting, lowering, pushing, pulling, and carrying tasks: Manual material handling guidelines. Retrieved May 5, 2017 from https://liberty mmhtables.libertymutual.com/CM_LMTablesWeb/pdf/LibertyMutualTables.pdf

Marras, W. S., Lavender, S. A., Leurgans, S. E., Rajulu, S. L., Allread, W. G., Fathallah, F. A., & Ferguson, S. A. (1993). The role of dynamic three-dimensional trunk motion in occupationally-related low back disorders: The effects of workplace factors, trunk position, and trunk motion characteristics on risk of injury. *Spine, 18*(5), 617–628.

Mattila, M., Karwowski, W., & Vilkki, M. (1993). Analysis of working postures in hammering tasks on building construction sites using the computerized OWAS method. *Applied Ergonomics, 24*(6), 405–412.

McAtamney, L., & Corlett, E. N. (1993). RULA: A survey method for the investigation of work-related upper limb disorders. *Applied Ergonomics, 24*(2), 91–99.

Neumann, W. P., Wells, R. P., & Norman, R. W. (1999). 4D WATBAK: Adapting research tools and epidemiological findings to software for easy application by industrial personnel. Paper presented at the *Proceedings of the International Conference on Computer-Aided Ergonomics and Safety*, Barcelona, Spain.

NIOSH. (1997). *Musculoskeletal disorders and workplace factors: A critical review of epidemiologic evidence for work-related musculoskeletal disorders of the neck, upper extremity, and low back* (No. 97-141). Bernard, B. P. (Ed.). Cincinnati, OH: U.S. Department of Health and Human Services, Public Health Service, Centers for Disease Control and Prevention, National Institute for Occupational Safety and Health.

NIOSH. (2016). New NIOSH smart phone app addresses ladder safety. Retrieved February 17, 2016, from http://www.cdc.gov/niosh/updates/upd-06-17-13.html

Occhipinti, E. (1998). OCRA: A concise index for the assessment of exposure to repetitive movements of the upper limbs. *Ergonomics, 41*(9), 1290–1311.

OSHA. (2015). Ergonomics eTool: Solutions for electrical contractors; Material handling—Heavy lifting. Retrieved September 25, 2015, from https://www.osha.gov/SLTC/etools/electricalcontractors/materials/heavy.html

Pheasant, S., & Haslegrave, C. M. (2006). *Bodyspace—Anthropometry, ergonomics and the design of work.* Boca Raton, FL: Taylor and Francis Group.

Siemens PLM Software. (2016). Jack and Process Simulate Humans. Retrieved August 27, 2016, from https://www.plm.automation.siemens.com/en_us/products/tecnomatix/manufacturing-simulation/human-ergonomics/jack.shtml

Silverstein, B. A., Fine, L. J., & Armstrong, T. J. (1987). Occupational factors and carpal tunnel syndrome. *American Journal of Industrial Medicine, 11*, 343–358.

Simple But Needed. (2016). Ergonomics evaluation. Retrieved February 17, 2016, from http://sbnsoftware.com/ergonomic-evaluation/

Snook, S. H., Vaillancourt, D. R., Ciriello, V. M., & Webster, B. S. (1995). Psychophysical studies of repetitive wrist flexion and extension. *Ergonomics, 38*(7), 1488–1507.

Stand up. (2016). Ergonomics by Stand up apps. Retrieved February 17, 2016, from http://www.ergonomicsapp.com/

Takala, E. P., Pehkonen, I., Forsman, M., Hansson, G. Å., Mathiassen, S. E., Neumann, W. P., Sjøgaard, G., Veiersted, K. B., Westgaard, R. H., & Winkel, J. (2010). Systematic evaluation of observational methods assessing biomechanical exposures at work. *Scandinavian Journal of Work, Environment & Health, 36*(1), 3–24.

3M. (2016). 3M safety apps. Retrieved February 17, 2016, from http://solutions.3m.com/wps/portal/3M/en_US/3M-PPE-Safety-Solutions/Personal-Protective-Equipment/safety-management/safety-training/3M-safety-Apps/

Waters, T. R., Putz-Anderson, V., Garg, A., & Fine, L. J. (1993). Revised NIOSH equation for the design and evaluation of manual lifting tasks. *Ergonomics, 36*(7), 749–776.

Workplace Safety & Prevention Services [WSPS]. (2011). Performing a physical demands analysis. Retrieved August 27, 2016, from http://www.wsps.ca/WSPS/media/Site/Resources/Downloads/Prfrmng_a_Physcl_Dmnds_Anlys_Instrctns_FINAL.pdf?ext=.pdf

## KEY TERMS

**anthropometry:** science that determine measures of people's size, shape, and functional capacity.

**biomechanics:** study of the human body from a mechanical standpoint; it refers to the study of strength, forces, and movement of muscles and limbs.

**ergonomics:** study of the relationship between the physical work environment and workers.

**standard operating procedures:** specific steps required to complete a task.

**work task demand:** requirements imposed by the work task on the person performing it.

**worker capacity:** what a worker's body can handle and produce—that is, its characteristics, including limitations, and capabilities or abilities.

**workspace design:** includes direct elements that workers interact with (e.g., tools, workbench, etc.) and indirect elements that workers do not interact with (e.g., lighting, sound, air quality, etc.).

# 9 Training Design

*Joseph R. Keebler, Elizabeth H. Lazzara,*
*and Deborah DiazGranados*

## CONTENTS

## INTRODUCTION

Training is integral to the success of almost all modern day organizations and must be adaptable to every learnable task and procedure demanded of various professions. The science of training is a multidisciplinary field, but with the advent of advanced technologies used for learning, the need for integrating training with human factors is greater than ever before. Therefore, this chapter will aim to provide an overview that touches on the major areas of training that can be aided by human factors practitioners and researchers. This chapter will review learning frameworks, training methods, research trends, applications, and future research within the context of human factors by providing an overview of the modern day conceptualization of training. First, we will review fundamentals, including establishing the definition of learning and current frameworks for training. Following, we will review methods for conducting valid training research, with a focus on designs that lead to the most valid results. We will then review training types and their various applications, followed by future trends based on emergent training technologies.

## FUNDAMENTALS

Learning, at its most basic level, is the process of acquiring, revising, or refining attitudes, behaviors, or cognitions. Ideally, learning eventually results in expert performance, which is sustaining superior performance in specific tasks within a particular domain (Ericsson, 2014). As shown in Figure 9.1, researchers present the complexities of learning by proposing frameworks depicting cognitive-based, behavioral-based, and affective-based outcomes (e.g., Bloom, 1956; Kraiger, Ford, & Salas, 1993). Cognitive-based outcomes include knowledge, knowledge organization, and cognitive strategies. *Knowledge* refers to declarative (knowledge about what), procedural- (knowledge about how), and strategic-knowledge (knowledge about which, when, and why; Wilson et al., 2009). Such knowledge is essentially the foundation and "building blocks" for learners. As learners progress through knowledge acquisition, they concurrently develop *knowledge organization* (i.e., meaningful structures for grouping information). Some argue that the way in which knowledge is organized is equally important to the amount of knowledge acquired (Johnson-Laird, 1983; Rouse & Morris, 1986). As novice learners become expert learners, their knowledge structure becomes more complex forming more intricate relationships (Glaser & Chi, 1988). In addition to knowledge and knowledge organization, learners develop cognitive strategies. *Cognitive strategies* are the techniques learners draw upon to plan, monitor, regulate, and solve problems, particularly in novel situations (Gagne, 1984).

Behavioral-based outcomes consist of compilation and automaticity. *Compilation* skills are attributable to composition (i.e., linking successive procedures into complex production) and proceduralization (i.e., combining integrated behaviors into an

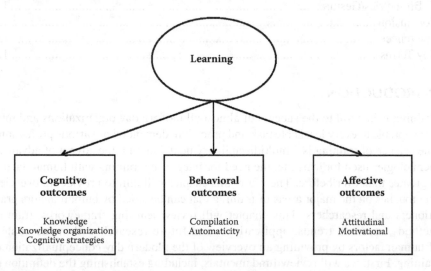

**FIGURE 9.1**  Cognitive, behavioral, and affective learning outcomes.

integrated routine; Kraiger, Ford, & Salas, 1993). It should be noted, though, that proceduralization is distinct from procedural knowledge as procedural knowledge is simply knowing how to perform a given task and proceduralization is synthesizing behaviors seamlessly. With compilation skills, learners can leverage previously acquired behaviors and apply them to novel situations. Meanwhile, *automaticity* is performing skills without controlled, mental processing. For example, most adults have automaticity when reading as it requires little to no controlled processing. Such level of performance enables learners to cope with extraneous demands since automaticity reduces the cognitive resources required (Brown & Bennett, 2002; Laberge & Samuels, 1974).

Finally, affective-based outcomes involve both attitudinal and motivational learning. Attitudinal outcomes reflect changes in internal states and preferences, whereas motivational outcomes include motivational dispositions, self-efficacy, and goal setting (Wilson et al., 2009). Although often overlooked, affective-based outcomes can be important, targeted end points as well as drivers of behavioral change.

## TRAINING EFFECTIVENESS FRAMEWORKS

Training effectiveness is a macrolevel, theoretical approach to understanding training-related outcomes by examining the entire system. Formally defined, training effectiveness is "the study of the individual, training, and organizational characteristics that influence the training process before, during, and after training" (Alvarez, Salas, & Garofano, 2004, p. 389). Essentially, training effectiveness entails discovering *why* training did or did not work effectively.

Researchers propose and examine these three types of characteristics: individual, organizational, and training, to uncover the aspects that contribute to training effectiveness (Goldstein & Ford, 2002). *Individual* characteristics refer to the factors that trainees contribute (e.g., personality, demographics, and expectations). *Organizational* characteristics pertain to the context of the implemented training (e.g., organizational climate, trainee selection process, and work policies). Finally, *training* characteristics involve the actual aspects of the training program (e.g., instructional style and feedback).

All of these aforementioned characteristics contribute to training effectiveness as indicated by reactions, posttraining attitudes, cognitive-based learning, training performance, transfer performance, and results. *Reactions* refer to the extent that trainees perceived the training to be enjoyable, useful, or relevant (Kraiger, 2002). *Posttraining attitudes*, as the name would suggest, are the attitudes that change as a result of training (e.g., motivation and self-efficacy). *Cognitive-based learning* is the acquisition of knowledge. *Training performance* is the ability to exhibit the acquired skills during training. Meanwhile, *transfer performance* is the ability to demonstrate the acquired skills in the work environment. *Results* are quantifiable changes related to trainees' transfer performance (e.g., quantity or quality of outputs; Tannenbaum, Cannon-Bowers, Salas, & Mathieu, 1993). See Table 9.1 for a summary of contributors to training effectiveness.

## CONSIDERATIONS FOR TRAINING DESIGN, IMPLEMENTATION, AND EVALUATION

Because training effectiveness is complex and multifaceted and learning is not a guarantee, it is imperative that it adhere to scientific principles of learning. To

## TABLE 9.1
## Contributors to Training Effectiveness

| Level of Characteristic | Type of Characteristic |
|---|---|
| Individual | • Personality traits |
| | • Attitudes |
| | • Abilities |
| | • Demographics |
| | • Experience |
| | • Expectations |
| Organizational | • Climate for learning |
| | • History |
| | • Policies |
| | • Trainee selection |
| | • Technique |
| | • Trainee notification |
| | • Process |
| Training | • Instructional style |
| | • Practice |
| | • Feedback |

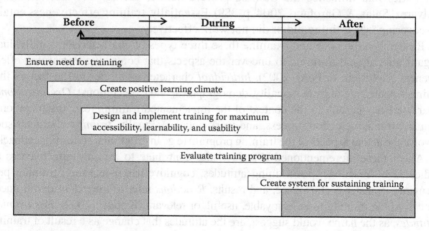

**FIGURE 9.2** Considerations for training programs. (Adapted from Salas, E., Benishek, L., Coultas, C., Dietz, A., Grossman, R., Lazzara, E., & Oglesby, J., *Team training essentials: A research-based guide.* New York: Routledge, 2015.)

maximize effectiveness, training should not be conceptualized as a one-time event. Rather, it should be conceptualized systematically beyond the actual training episode. As illustrated in Figure 9.2, Salas et al. (2015) propose five pillars that serve as the scientific foundation of training:

1. Ensure need for training
2. Create positive learning climate

3. Design and implement training for maximum accessibility, learnability, and usability
4. Evaluate training program
5. Create a system for sustaining training

The heart of Pillar 1 is establishing the need for the training itself as well as the targeted cognitions, behaviors, and attitudes. This necessity is proven by conducting a needs analysis, which identifies the organizational resources required to conduct training, the task activities necessary to perform the job, and the individuals who need training (Goldstein & Ford, 2002). Pillar 2 focuses on preparing the learner and learning environment by framing training positively, establishing a supportive, encouraging learning atmosphere, and facilitating learners' motivations, efficacy, and engagement. Pillar 3 concentrates on the actual training design. An efficacious training is created through partnerships between training and subject matter experts, identifies learning objectives in advance, and implements multi-method instructional strategies (i.e., information, demonstration, practice, and feedback). Practice-based strategies, deliberate practice in particular, as well as feedback are especially important as they are foundational for acquiring accurate performance and expertise (Ericsson, Krampe, & Tesch-Romer, 1993). Such strategies involve the use of explicit instructions regarding methodologies, supervision to diagnose errors, and informative feedback. Pillar 4 centers around evaluating the training program by employing a multilevel evaluation approach (e.g., Kirkpatrick, 2004) and using the evaluative data to refine future training iterations. Finally, pillar five centralizes on establishing a system for sustaining the training program as well as the trained attitudes, behaviors, and cognitions by facilitating practice opportunities and transfer. Sustainment efforts include strategies such as modeling behavior (Taylor, Russ-Eft, & Chan, 2005), providing practice opportunities (Salas, Tannenbaum, Kraiger, & Smith-Jentsch, 2012), and establishing rewards and incentives (Epstein, 2008). Such sustainment efforts are crucial to ameliorate decay, which is probable considering that as much as 92% of trained KSAs deteriorate within a year of initial training (Arthur, Bennett, Stanush, & McNelly, 1998).

## METHODS

### RESEARCH DESIGN WITHIN THE TRAINING CONTEXT

#### Considerations for Establishing the Criterion

Training design, validation, and implementation require precise measurement of what are often called *criterion measures* (Goldstein & Ford, 2002), which are the outcome metrics used to determine the efficacy and success of a training program. These measures further include the concepts of *criterion relevancy*, or the degree of importance a particular criterion has in regard to needed knowledge, skills, and attitudes (KSAs); *criterion deficiency*, or the degree to which the criterion lacks specificity; *criterion contamination*, or the presence of extraneous or third variables that can lead to changes in the criterion unrelated to training; and *criterion reliability*, or the degree to which the measure of the criterion is consistent (Goldstein & Ford, 2002). Keeping the criterion in mind, along with the elements stated previously that can affect one's ability to interpret whether training was valid, is the basis for strong training research designs.

## Pretesting, Posttesting, and Control Groups

All training studies should implement pretesting of the required KSAs, followed by multiple posttests. One posttest immediately following training will not suffice to understand training validity (Goldstein & Ford, 2002) and, in general, is not a very powerful research design (Shadish, Cooke, & Campbell, 2002). Further, training studies should utilize a control group when possible. Utilization of a control allows for comparison of the trained group to a non-trained group, allowing the researcher to dissect potential confounding effects, that is, the effect of the passage of time, or the effect of learning happening in the organization that is not part of the training program.

## POTENTIAL PITFALLS OF TRAINING EVALUATION

Oftentimes, organizations are constrained by time and resources in regard to implementing and evaluating training. Given these constraints, there are issues that must be addressed to ensure that training is delivered and evaluated in the most efficacious manner. This section will briefly touch on some of the major issues that can arise from poor training evaluation strategies, with a particular focus on lack of a control group, lack of a pretest, or both. It is almost always better to include both a control group and a pretest. Control groups provide a baseline representation of the population of interest, allowing us a rational basis for comparison if our treatment works. Pretests allow us to avoid individual differences between participants that could cause an effect, leading us to mistakenly assume our treatment works when in fact it is actually an individual variable. Regardless of the presence or absence of either a control group or pretest, there are instances where causal inferences can still be made when one or both are missing from a particular study.

## Studies with No Control Group

Studies without a control group suffer from an inability to know whether the training had a causal effect on the outcomes. Specifically, the researcher would be unable to disentangle whether the results were caused by the training or caused by some predisposition or quality of the group being trained. The best way to remedy this effect is through a *repeated-treatment design with a removed treatment*. This is usually described using Os for *observations* and Xs for *treatments*, as follows: $O_1 \, X \, O_2 \, \cancel{X} \, O_3 \, X \, O_4$ (Shadish et al., 2002). The pattern of results for this design should be that $O_2 > O_1$ (effect of treatment), $O_3 < O_2$ (after treatment was removed), and $O_4 > O_3$ (after the treatment was re-instituted). If this relationship maintains, then the results provide compelling evidence that the training is what is causing the effect. Researchers must be wary that seasonal or cyclical differences, such as time of year or day of the week could have potential impacts on, say, productivity.

## Studies with No Pretest

The second type of study that can cloud causal assumptions is the study with a control group but no pretest. Although pretesting can have influential effects on the outcomes, whereas some individuals learn from the pretest and therefore perform better on the posttest solely for that reason, in general, introducing a pretest can help with causal explanation. One of the best designs for studies lacking a pretest is the *posttest-only design with proxy pretests*. This design implements some measure related to the posttest that has been assessed previously for the group of interest.

For instance, if the organization wants to understand effects of training on email response efficiency, they might use emails from within the company to get an understanding of each individual's ability to communicate prior to the experiment. Maybe the organization is interested in training that decreases the amount of time it takes for individuals to respond to emails. If this were the case, the company could analyze previous emails sent over the last year for each participant and get an average response time. This could then be used as a proxy for a pretest related to the post-tested variable and would allow the researcher's to understand whether the training was the cause of a reduction in response time, or whether it was simply that individuals who were trained happened to be those who responded quickly to emails in the first place. Ideally, we would like to see a mixture of fast responders with slow responders, and that the training significantly decreases the time that both groups take to respond after training has been implemented.

## TRAINING TYPES

An instructional approach is comprised of the tools used to develop the training, the methods used to deliver the training, and the content of the training (see Figure 9.3; Salas & Cannon-Bowers, 1997, 2001). While no one instructional approach fits every training need, the science of training continues to expand and adapt because new approaches are being developed. Due to the nature of work (e.g., team based and complex, high stakes), some of the most widely researched instructional approaches are team-based training and synthetic learning environment (SLE) based training

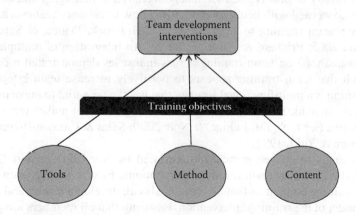

**FIGURE 9.3**   Team Development Intervention Model. (Adapted from Salas, E., & Cannon-Bowers, J. A., Methods, tools, and strategies for team training, in M.A. Quinones & A. Ehrenstein (Eds.), *Training for a rapidly changing workplace: Applications of psychological research* (pp. 249–279). Washington, DC, American Psychological Association, 1997.)

(i.e., simulation-based or games used for training). Before discussing these two broad training topics, in addition to understanding the five pillars that represent the scientific foundation of training (Salas et al., 2015), it is important to further delineate the principles of training and how this may apply to team training. Noe and Colquitt (2002) highlighted seven principles of effective and well-designed training: (1) trainees understand the objectives, purpose, and intended outcomes of the training; (2) the content, examples, exercises, and assignments are meaningful and relevant to the trainees; (3) trainees are provided with learning aids to help them learn, organize, and recall content; (4) the practice environment is a safe environment; (5) trainees receive feedback; (6) trainees can observe and interact with other trainees; and (7) the training is coordinated effectively. These principles focus on the factors (e.g., individual, environment, climate, organizational, content) that influence training effectiveness.

A key finding within the training research literature is the effectiveness of providing opportunities for practice and providing feedback (Kluger & DeNisi, 1996; Schmidt & Bjork, 1992; Schmidt & Wulf, 1997). However, all practice and feedback are not created equal. Practice should be structured, it should take place in a safe environment where errors can be made and learned from, and the practice environment should be similar to the transfer environment. Feedback should be timely, specific, task focused and diagnostic (Cannon & Witherspoon, 2005; Kluger & DeNisi, 1996). Maximizing transfer of training is the goal for training developers. In the text that follows, the two overarching topics of team training and SLE based training will be discussed in terms of what makes team training and SLE based training most effective.

## Team Training

Teams are prevalent in all organizations. Individuals work in teams of different sizes; they work in multiple teams and have different roles in each team and they conduct a variety of task types. Individuals in charge of managing and developing teams are concerned with better understanding how to enhance teamwork. Studies have shown team training to be effective (e.g., Brannick, Prince, & Salas, 2005; Ellis et al., 2005; Prichard & Ashleigh, 2008). An integration of multiple studies (i.e., meta-analysis) on team training across industries demonstrated a consistent effect such that team training appears to positively increase team effectiveness. Team training is a multifunctional strategy that has the potential to make multilevel impact (e.g., individual, team, and organizational). But what makes team training most effective (see Colquitt, LePine, & Noe, 2000; Salas & Cannon-Bowers, 2001; Tannenbaum & Yukl, 1992)?

Let us apply the seven principles delineated by Noe and Colquitt (2002) to highlight the needs for team training interventions. First, effective team training provides training content so that all individuals can develop a shared and accurate mental model of the training intervention. Ensuring that all members are prepared for the training and understand the need for their participation in the training as well as its intended outcomes is critical to developing effective team training. Second, team training is not a panacea; in other words, it will not solve all team relevant process and performance problems. The content and activities must be relevant to the trainees. Therefore, the team must be able to link the team training to their particular role on the team and within the organization. Third, learning aids are provided to all team members, not only to aid them during training but to also ensure the transfer of training. To build team-based procedural knowledge and team level cognitive strategies, effective team training should provide aids for team members to understand their role within the team and its integration with the other members of the team. Fourth, the training environment is safe for all team members. Team dynamics (e.g., leadership/followership) should not prohibit the training environment to be safe for individuals to practice, make mistakes, and develop their KSAs during training. Fifth, all team members should receive feedback during training. Feedback enhances the retention process and is critical to effective team training (Schmidt & Bjork, 1992). However, the use of feedback and its timing is an important consideration. Too frequent feedback can block information processing activities that are necessary during the acquisition phase of training, but not enough feedback may not provide enough information to the team to provide process, learning, and performance indicators during training. Sixth, team training should be provided to the team as a unit. Effective team training encourages team members to interact and develop their skills as a unit. Seventh, the coordination of team training cannot be overlooked nor its effect on team training effectiveness. Team training interventions are difficult to coordinate. Individuals charged with developing team training within organizations are often required to coordinate individuals across departments or functions. Effective coordination of schedules and locations is required in order to ensure that team training is executed without distractions.

## SLE-Based Training

SLE-based training is a training approach developed that focuses on providing trainees with opportunities to observe, practice, and develop the required competencies and receive feedback in a safe environment (Slotte & Herbert, 2008). SLE-based training provides the opportunities to practice and work real-world problems in work-related scenarios (Ng & Ng, 2004). Simulated opportunities also provide the trainees with time to reflect on their performance and how the knowledge and skills can be used on the job (Brown, Collins, & Duguid, 1989). To create effective SLE-based training, interventions must adhere to the foundations of all training interventions (e.g., needs analysis, developing learning objectives, providing feedback). However, additional measures must be taken when developing SLE-based training to ensure its effectiveness. In particular, great detail and work should be attributed to the development of the scenarios used during the training as well as the feedback provided to the trainees. Simulation scenarios must be developed to appropriately initiate the desired competencies. For example, if SLE-based training is targeted to develop communication skills, the training should be designed so that team members interact with one another. Feedback is critical in SLE-based training since the focus of this training approach is to provide trainees with opportunities to practice. The development of effective behavioral performance measures is critical for providing effective feedback to trainees. Behavioral performance measures include automated performance measures such as those gathered by the simulation (e.g., time to determine heart failure in a medical simulator, or angle of approach). Other behavioral metrics may include observational ratings or process measures in which observers rate how the team responds to the events that are meant to initi- ate the desired competencies. The development of the metrics and the training of observers are equally critical in ensuring the effectiveness of an SLE-based training intervention.

## APPLICATION

The beginning of this chapter focused on training design and the major principles that affect the development and evaluation of training. This next section will illus- trate the application of these concepts.

### Case Study: U.S. Army Trauma Training Center and the Five Pillars

The U.S. Army Trauma Training Center's (ATTC) mission provides a strong example for the five pillars. Pillar 1, "Ensure a need for training": The U.S. ATTC is respon- sible for preparing Forward Surgical Teams (FSTs), a 20-person medical team man- aging emergent trauma scenarios on the battlefield, for deployment. Pillar 2, "Create a positive learning climate": To prepare for this mission, these FSTs undergo a rig- orous, two-week training prior to deployment that is rooted in established task and team competencies. Pillar 3, "Design and implement training for maximum acces- sibility, learnability, and usability": The training is delivered with multiple instruc- tional strategies (information, demonstration, practice, and feedback instructional strategies) over the course of three phases. During Phase 1, FSTs receive didactic instruction and participate in various simulations. FSTs practice specific skills with part-task trainers, manage trauma cases on patient simulators, and attend trauma and surgical courses. During Phase 2, FSTs participate in clinical rotations at the trauma resuscitation unit, operating room, and trauma intensive care unit to care for live patients. Phase 3 culminates in a capstone exercise where FSTs assume responsibil- ity of the trauma resuscitation unit, observation area, and trauma operating room, as this most closely resembles what FSTs would experience during a mass casualty event. Pillar 4, "Evaluate training program": Throughout each of the previous three phases, the team is evaluated through testing and expert observation to ensure the appropriate KSAs are being exhibited and learned. Pillar 5, "Create a system for sustaining training": This training program is a required and valued aspect of the Army's ATTC and has provided an excellent standard for preparing FSTs since the 1980s.

## FUTURE TRENDS

A multitude of emergent technologies can potentially enhance both the trainee expe- rience and training outcomes. Next, we will discuss two of these in relation to the topics introduced in this chapter.

## Augmented Reality

Augmented reality (AR) is a type of mixed-virtual environment that overlays computer-generated imagery onto real-world visual scenes. AR differs from virtual reality in that it overlays digital information onto a real-world scene vs. virtual reality where the entire visual scene is digitally created. Within the context of training, AR can provide numerous benefits to the training environment, including real-time feedback on procedures and tasks; real-time feedback in regard to trainee performance; trainee insight into expert mental models; and overlay of relevant information without forcing the trainee to cross-reference a text or instruction manual. Although it has been around for approximately two decades, due to prohibitive costs, AR has only recently become accessible to the wider consumer community. A considerable body of research on using AR for training has been relegated to the medical domain, where it has been used to teach aspects of surgery (Keebler, Patzer, Wiltshire, & Fiore, 2017).

## Biometrics/Gesture

The current model for education in healthcare, particularly in specialties such as surgery, may involve simulation-based training and years of practical experience to perfect skills such as arm-hand dexterity, finger dexterity, and control precision. Simulators provide a realistic and safe environment for surgical scenarios to be reproduced and repeated. The future of training, particularly in the healthcare industry, may lend itself to training that incorporates gesture training.

Gesture training includes training that utilizes technology that provides feedback to the trainee regarding the position and tracking of movements to improve task performance. Currently, a few technologies on the market have been used to train hand movements. For example, the Leap Motion controller and the Xbox Kinect have been used in initial research for training surgical skills and even in sign language training. This technology recognizes hand movements and translates them into changes of the position of the AR object. Once movements are recognized, feedback can be provided to trainees regarding their positioning and precision.

## CONCLUSION

This chapter consisted of a high-level overview of the current state of training design, implementation, and validation. Training continues to be one of the most effective methods for enhancing performance in the workplace, and with the advent of modern technologies, we will continue to see training evolve in ways unimaginable even a decade ago. Virtual technologies will allow for training programs to reach a wider audience of individuals in the workforce than in previous decades; AR systems will allow for on-the-job training where individuals can literally see what an expert sees; and biometrics will aid in the understanding of learned behaviors and skills. Human factors will continue to play a very important role to ensure that these training systems remain valid, safe, and user friendly.

# REFERENCES

Alvarez, K., Salas, E., & Garofano, C. M. (2004). An integrated model of training evaluation and effectiveness. *Human Resource Development Review, 3*(4), 385–416.

Arthur, W. Jr., Bennett, W. Jr., Stanush, P. L., & McNelly, T. L. (1998). Factors that influence skill decay and retention: A quantitative review and analysis. *Human Performance, 11*(1), 57–101.

Bloom, B. S. (Ed.). (1956). *Taxonomy of educational objectives, Handbook 1: Cognitive domain.* New York: David McKay.

Brannick, M. T., Prince, C., & Salas, E. (2005). Can PC-based systems enhance teamwork in the cockpit? *The International Journal of Aviation Psychology, 15*(2), 173–187.

Brown, S. W., & Bennett, E. D. (2002). The role of practice and automaticity in temporal and nontemporal dual-task performance. *Psychological Research, 66,* 80–89.

Brown, J. S., Collins, A., & Duguid, P. (1989). Situated cognition and the culture of learning. *Educational Researcher, 18*(1), 32–42.

Cannon, M. D., & Witherspoon, R. (2005). Actionable feedback: Unlocking the power of learning and development. *Academy of Management Executive, 19,* 120–134.

Colquitt, J. A., LePine, J. A., & Noe, R. A. (2000). Toward an integrative theory of training motivation: A meta-analytic path analysis of 20 years of research. *Journal of Applied Psychology, 85*(5), 678.

Ellis, A. P., Bell, B. S., Ployhart, R. E., Hollenbeck, J. R., & Ilgen, D. R. (2005). An evaluation of generic teamwork skills training with action teams: Effects on cognitive and skill-based outcomes. *Personnel Psychology, 58*(3), 641–672.

Epstein, M. J. (2008). Making sustainability work: Best practices in managing and measuring corporate, social, environmental, and economic impacts. Sheffield, United Kingdom: Greenleaf Publishing Limited.

Ericsson, K. A. (2014). Why expert performance is special and cannot be extrapolated from studies of performance in the general population: A response to criticisms. *Intelligence, 45,* 81–103.

Ericsson, K. A., Krampe, R. Th., & Tesch-Romer, C. (1993). The role of deliberate practice in the acquisition of expert performance. *Psychological Review, 100*(3), 363–406.

Gagne, R. M. (1984). Learning outcomes and their effects: Useful categories of human performance. *American Psychologist, 39*(4), 377–385.

Glaser, R., & Chi, M. T. H. (1988). Overview. In M. T. H. Chi, R. Glasert, & M. Farr (Eds.), *The nature of expertise* (pp. xv–xxviii). Hillsdale, NJ: Erlbaum.

Goldstein, I. L., & Ford, J. K. (2002). *Training in organizations.* Belmont, CA: Wadsworth.

Johnson-Laird, P. (1983). *Mental models.* Cambridge, MA: Harvard University Press.

Keebler, J. R., Patzer, B. S., Wiltshire, T., & Fiore, S. (2017). Augmented reality in training. *Cambridge Handbook of Training.* Manuscript in preparation.

Kirkpatrick, D. L. (2004). *Evaluating training programs: The four levels.* San Francisco: Berrett-Koehler.

Kluger, A. N., & DeNisi, A. (1996). The effects of feedback interventions on performance: A historical review, a meta-analysis, and a preliminary feedback intervention theory. *Psychological Bulletin, 119*(2), 254.

Kraiger, K. (2002). Decision-based evaluation. In K. Kraiger (Ed.), *Creating, implementing, and managing effective training and development* (pp. 331–375). San Francisco: Jossey Bass.

Kraiger, K., Ford, J., & Salas, E. (1993). Application of cognitive, skill-based, and affective theories of learning outcomes to new methods of training evaluation. *Journal of Applied Psychology, 78*(2), 311–328.

Laberge, D., & Samuels, S. J. (1974). Toward a theory of automatic information processing in reading. *Cognitive Psychology, 6,* 293–323.

Ng, D. F. S., & Ng, P. T. (2004). Computer simulations: A new learning environment for professional development of educational leaders. *Educational Technology, 44*(6), 58–60.

Noe, R. A., & Colquitt, J. A. (2002). Planning for training impact: Principles of training effectiveness. In K. Kraiger (Ed.), *Creating, implementing, and maintaining effective training and development: State-of-the-art lessons for practice* (pp. 53–79). San Francisco: Jossey-Bass.

Prichard, J. S., & Ashleigh, M. J. (2007). The effects of team-skills training on transactive memory and performance. *Small Group Research, 38*(6), 696–726.

Rouse, W. B., & Morris, N. M. (1986). On looking to the black box: Prospects and limits in the search for mental models. *Psychological Bulletin, 100,* 349–363.

Salas, E., Benishek, L., Coultas, C., Dietz, A., Grossman, R., Lazzara, E., & Oglesby, J. (2015). *Team training essentials: A research-based guide.* New York: Routledge.

Salas, E., & Cannon-Bowers, J. A. (1997). Methods, tools, and strategies for team training. In M. A. Quinones & A. Ehrenstein (Eds.), *Training for a rapidly changing workplace: Applications of psychological research* (pp. 249–279). Washington: American Psychological Association.

Salas, E., & Cannon-Bowers, J. A. (2001). The science of training: A decade of progress. *Annual Review of Psychology, 52*(1), 471–499.

Salas, E., Tannenbaum, S. I., Kraiger, K., & Smith-Jentsch, K. A. (2012). The science of training and development in organizations: What matters in practice. *Psychological Science in the Public Interest, 13*(2), 74–101.

Schmidt, R. A., & Bjork, R. A. (1992). New conceptualizations of practice: Common principles in three paradigms suggest new concepts for training. *Psychological Science, 3*(4), 207–217.

Schmidt, R. A., & Wulf, G. (1997). Continuous concurrent feedback degrades skill learning: Implications for training and simulation. *Human Factors, 39*(4), 509–525.

Shadish, W. R., Cook, T. D., & Campbell, D. T. (2002). *Experimental and quasi-experimental designs for generalized causal inference.* Belmont, CA: Wadsworth.

Slotte, V., & Herbert, A. (2008). Engaging workers in simulation-based e-learning. *Journal of Workplace Learning, 20*(3), 165–180.

Tannenbaum, S. I., Cannon-Bowers, J. A., Salas, E., & Mathieu, J. E. (1993). *Factors that influence training effectiveness: A conceptual model and longitudinal analysis* (Technical Report No. 93-011). Orlando, FL: Naval Training Systems Center.

Tannenbaum, S. I., & Yukl, G. (1992). Training and development in work organizations. *Annual Review of Psychology, 43*(1), 399–441.

Taylor, P. J., Russ-Eft, D. F., & Chan, D. W. (2005). A meta-analytic review of behavior modeling training. *Journal of Applied Psychology, 90*(4), 692–709.

Wilson, K. A., Bedwell, W. L., Lazzara, E. H., Salas, E., Burke, C. S., Estock, J., Estock, J. L., Orvis, K. L., & Conkey, C. (2009). Linking game attributes to learning outcomes: State of play and research paths. *Gaming & Simulation, 40*(2), 217–266.

## KEY TERMS

**cognitive-based learning:** acquisition of knowledge.

**criterion measures:** outcome metrics used to determine the efficacy and success of a training program.

**posttraining attitudes:** attitudes that change as a result of training.

**reactions:** extent that trainees perceived the training to be enjoyable, useful, or relevant.

**results:** quantifiable changes related to trainees transfer performance.

**synthetic learning environment (SLE)-based training:** training approach that provides trainees with opportunities to observe, practice, and develop the required competencies and receive feedback in a safe environment.

**team training:** set of tools and methods used to train a group of individuals on both task-related individual competencies and teamwork competencies.

**training effectiveness:** study of the individual, training, and organizational characteristics that influence the training process before, during, and after training.

**training performance:** ability to exhibit the acquired skills during training.

**transfer performance:** ability to demonstrate the acquired skills in the work environment.

# Section III

Putting Human Factors
into Practice

# 10 Looking Ahead: Human Factors in Sociotechnical Systems

*Dan Nathan-Roberts and David Schuster*

## CONTENTS

## A FRAMEWORK FOR THE FIELD

A useful method for understanding the field of human factors is to think of it as a ladder. The first step (rung) on the ladder begins with an individual physically pushing and pulling things in a physical environment. The next steps focus on how an individual knows what and how to push and pull, the dynamics of a team pushing and pulling, and the impact of an organization on the rungs below it. At the highest level is how society is structured to encourage the formation of the rungs and the shape of the ladder. The human-tech ladder (Figure 10.1) is a good model for understanding human factors as a series of areas of research that have their origins in an individual's physical needs.

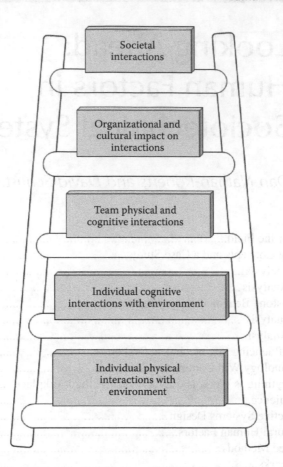

**FIGURE 10.1** Overview of human factors as a ladder. (Adapted from the Human-Tech Ladder proposed by Vincente, K., *The human factor: Revolutionizing the way people live with technology.* London, UK, Routledge, 2004.)

The human-tech ladder is a useful tool with which to cement understanding of how the various areas of human factors are different, and it is also important to tying the various subdisciplines together. It is rare that an ergonomic problem occurs in a vacuum. Most large ergonomic problems come from the confluence of a variety of factors in combination. Fixing these problems can be difficult, but doing so also provides the opportunity for some of the biggest payoffs. To solve these types of problems, ergonomists use sociotechnical system models to identify how all of the aspects of the ladder above work together simultaneously while also looking at each of the rungs of the ladders: in other words, seeing the forest *and* the trees.

Sociotechnical models are frameworks to organize the various subdisciplines and their relationships to each other and the outcome. Figure 10.2 shows a sociotechnical model that ties various subdisciplines together. One of the commonalities across sociotechnical systems models or frameworks is that they cross many subdisciplines of human factors. The key takeaway from sociotechnical systems is that opportunities

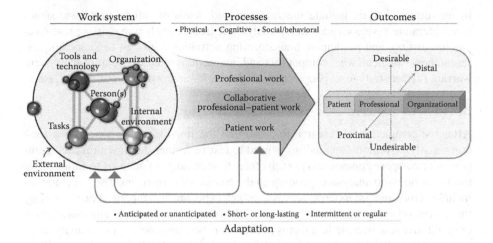

**FIGURE 10.2** SEIPS 2.0, another sociotechnical systems model. (From R.J. Holden et al., *Ergonomics,* 56(11), 1669–1686, 2013. With permission.) (For more examples, refer to Van Houdt, Heyrman, Vanhaecht, Sermeus, & De Lepeleire, 2013.)

for improvement often do not fit squarely in just one discipline of human factors. By simultaneously considering multiple disciplines and their interactions, we can make the largest improvement. It is important to improve all these factors rather than striving for perfect, single-discipline optimization. Like many things in life, successful implementation is about making choices and trade-offs. When applying human factors concepts to real problems, it is important to keep the larger system in mind so that you can proactively determine the areas and methods that will have the largest impact.

In the same manner that sociotechnical systems are useful in solving problems that we can see, sociotechnical systems are also useful in directing our attention to unknown problems. This is also known as top–down analysis. Top–down analysis is used when we do not yet know where to focus our attention; in other words, we do not know what we are looking for. Studying small components of the system and their interactions is called bottom–up analysis, which has the benefit of solving a particular problem. Sociotechnical models can be used as a guide to components of the system and how components interact. Because of this, they can lead you to adopt additional perspectives and proactively resolve problems.

## TOOLS FOR UNITING CONCEPTS AND A CASE STUDY

In this section, we provide a concrete example of simultaneous high- and low-level analysis as well as top–down and bottom–up analysis. Note that some research techniques can be used for bottom–up or for top–down analysis; it is the goal or research question that differs in each scenario.

### TOP–DOWN ANALYSIS

Top–down analysis is a set of techniques that allow researchers to understand what components and variables have an impact on the final outcomes. Techniques used in top–down analysis include unstructured and semistructured interviews, struc-tured literature reviews, quantitative analysis of trends, analyzing existing text (cor-pus linguistics), and participant brainstorming activities. Through these techniques, researchers can glean what components and interactions of components of the system warrant further study (e.g., people, tools, tasks, task–environment components, etc.).

## BOTTOM–UP ANALYSIS

After the completion of research to determine the important components via top–down analysis, bottom–up analysis is used to test in further detail within each com-ponent or compare components to each other. A wide range of analysis techniques are used in bottom–up analysis depending on the components involved. These techniques include structured interviews, surveys, in-lab experiments, forced choice ranking, three-dimensional postural analysis, tests of workload, and almost any quantitative or qualitative research method that is relevant for the component(s) being analyzed.

Looking to Wickens, Lee, Liu, and Gordon-Becker (2003) for further clarifica-tion, consider a comparison between the two types of research. Top–down research seeks to identify what items of the system should be in the model. In contrast, bottom–up seeks to identify the fine nuances between and within each item. For example, a top–down research method would be used to determine if scissors used for a spe-cific surgery should have a locking mechanism (to keep them closed when not in use), whereas bottom–up research would be used to determine the optimal locking mechanism and exact position of the lock for easy use.

## CASE STUDY: KEYSTONE BEACON COMMUNITY PROJECT

A good example of an in-depth sociotechnical system analysis was completed on the impact of the Keystone Beacon Community (KBC) performed by the University of Wisconsin-Madison's Center for Quality and Productivity Improvement (CQPI). The KBC is a multiyear project aiming to improve healthcare coordination within and between healthcare providers in Central Pennsylvania. Led by the Geisinger Health System and funded by the Office of the National Coordinator for Health IT, KBC's goals included lowering hospital admissions and readmissions for patients with chronic obstructive pulmonary disease and other chronic diseases.

CQPI performed a multiyear in-depth sociotechnical analysis of the facilitators and barriers to care from the project implementation. This analysis used observations, interviews, focus groups, longitudinal analysis, electronic healthcare records, and surveys to collect data. The data were collected from many of the groups involved, including patients, care managers, clinicians, the care coordination team, and the project management team. The analysis sought to understand, at a high level, all of the components of the system in Figure 10.2 and more deeply investigate the com-ponents that most heavily impacted the patient care outcomes, such as readmissions.

### TOP–DOWN ANALYSIS

To perform a top–down analysis of the patients, several techniques were used in con-junction with 19 participants. The participants were recruited for structured face-to-face interviews, focus groups, and telephone interviews. Patients were screened to fit a set of inclusion criteria (e.g., recently receiving care from the program, no diagnosis of altered mental state, etc.). During the structured interviews, quantitative subjective questions were asked, including "How would you rate your care on a scale from one to five?" Unstructured qualitative questions were also asked, such as "How has your experience with a care manager been?" This approach, also sometimes called a mixed-methods approach, was used to simultaneously collect detailed information that could be compared across participants, such as average quality of care received, as well as qualitative data that could help explain the findings and put them in a larger context, such as why someone might rate their care higher or lower than previous experiences. From these qualitative and quantitative techniques, researchers learned about the complex array of comorbidities in patients (multiple diseases such as diabetes and congestive heart failure), the most useful tasks performed by care managers, and how memorable the in-patient and out-patient care managers were.

## BOTTOM–UP ANALYSIS

The data found in the top–down analysis were used to inform bottom–up experiment design and analysis that consisted of a paper survey sent to a large number of patients who had received care with and without care managers at several health institutions. The survey allowed the researchers to further hone in on topics uncovered by the top–down analysis, compare the relative importance of topics discovered in the mixed-methods study, compare patients who had and had not received care from a care manager, and reach a much larger audience with fewer resources. The survey asked for information about the patient's health, patient activation (Hibbard, Mahoney, Stockard, & Tusler, 2005), and services provided by care managers, including patient education, medication reconciliation, patient-specific action plans, and referrals.

One of the findings of the top–down analysis was that it was difficult to remember the care managers specifically. Participants were overwhelmed by a large number of care providers in the hospital (in-patient providers), making it difficult to remember any individual person. In comparison, each patient had fewer outpatient providers, making it seem more likely that patients would remember care managers. This hypothesis (that outpatient care managers were more memorable than in-patient care managers) was able to be tested by the bottom–up survey in a more robust manner than would have been possible in focus groups or interviews.

Identical surveys were sent to 404 patients (204 of whom had received care from a care manager). Surveys were received back from both groups (78 from patients who received care management and 88 who were treated at similar hospitals but did not have care managers). The bottom–up survey found that only 29% of patients who were treated by a care manager remembered the care manager, while 26% who weren't treated by a care manager thought they were. However, 54% of participants who received care from an outpatient care manager recalled that care manager, compared to just 36% of the non-care-manager group who incorrectly said they remembered having a care manager. In this case, the bottom–up analysis was better suited to identifying subtle differences between the two groups.

In addition to collecting more-nuanced details about the participants, the survey allowed the researchers to reach a larger participant pool, such as participants who were less-mobile and those who were unable to come in without drastically increasing the cost to complete the study. Completed surveys were coded (i.e., yes = 1, no = 2, illegible = 3, blank = 4) and entered into a database by multiple researchers. The standardized coding and multiple researcher data entry method caught errors in the data entry that might have been missed if only one person entered the data.

The combination of top–down and bottom–up mixed-methods studies that were performed across the sociotechnical system allowed researchers to identify components and relationships between components in the sociotechnical system that were especially high impact and then to study these results in a way that was more detailed than either method could have produced by itself. In other top–down and bottom–up sociotechnical analyses, the bottom–up analysis method may include more controlled studies, such as laboratory tests of small subsets of the system.

## FUTURE TRENDS INFLUENCING SOCIOTECHNICAL SYSTEMS

In this next section, we briefly discuss six trends we have observed in our field or in society that will impact human factors in the future both near and far. From technology to methodology, we discuss many potential upcoming changes.

### TREND 1: TECHNOLOGY WILL CONTINUE TO DISRUPT

The digital revolution has been the greatest change to the physical and cognitive world of work since the Industrial Revolution. Computers have allowed us to accomplish more by automating information processing. Today, computers and robots perform countless dull, dirty, and dangerous tasks that previously occupied much of the workday for many people. At the same time, new challenges have arisen as barriers to the amount of information that can be processed and the speed at which people can communicate have all but vanished. Human factors have been well positioned to improve safety, efficiency, and enjoyment when humans and technology work together.

The progress of computer technology can be extrapolated to the future. Although the rate of improvement is uncertain, computers will continue to improve in processing speed and storage capacity, and they will become increasingly smaller. As computers have become smaller, mobile devices have emerged. As this trend continues, the Internet of Things (defined as Internet connectivity between everyday objects; New Oxford American Dictionary, 2016) will become an increasing part of daily life; devices will get smarter as many more artifacts, from coffee pots to cars, become connected to the Internet. For example, smart homes can tailor lighting, appliances, heating and cooling, and security monitoring to their occupants' behavior rather than being explicitly commanded. If done well, the resulting system would place fewer cognitive and temporal demands on users while using energy more efficiently than is possible through manual control.

The world of work will also continue to be disrupted by new technology. This can be seen as an increasing dependence on increasingly sophisticated software. Tools and tasks that have been situated squarely in the physical world are turning into software. Banking is an example where the last remaining paper-based components are now digitized. For customers, this means that physical trips to the bank to process paperwork have ceased, becoming completed solely by software. This trend will continue for other things we think of as inherently physical today. From visiting the library to grocery shopping, technology disrupts by removing some barriers to productivity and introducing new challenges to safety, efficiency, and satisfaction. For example, cybersecurity and digital privacy are now issues of physical security and safety as much as they are issues in information security. The potential for increases in productivity, safety, and user satisfaction will require a sophisticated and pervasive science to understand how humans interact with these disruptive technologies. Human factors uniquely addresses these challenges.

## TREND 2: EVERYTHING IS WORK, AND EVERYWHERE IS THE WORKPLACE

Humans are continually looking for ways to do more and become more efficient. Utopian dreams of only working 15 hours a week (because machines are doing everything) have been usurped by the technology advancement treadmill and products as services (e.g., Pandora, cloud storage, etc.). To enable this, work is taking more forms and happening in more locations. Smart devices have enabled us to write, code, answer e-mails, edit spreadsheets, and complete any other number of tasks from almost anywhere on the planet (as witnessed by the fact that this chapter was completed thanks in part to electronic communication and without any of its contributors working together in the same room). Moving forward, the workweek will continue to blur with the non-workweek for many. The increase in job turnover means that employees will find themselves working in different corporate cultures, too. We are seeing the pinnacle of this manifestation right now with the regrowth of an on-demand workforce enabled by Uber, Task Rabbit, Mechanical Turk, and other similar services.

Why does this matter? In short, to maximize the utility of this mobile work force, it is important that we design work such that it can be done on the go, from home, or even while using a subscription to a self-driving car service. Some companies are getting there even at the time of this writing; Delta Airlines, the world's largest air carrier, allows reservations employees to work from home. SalesForce, an online customer relationship management software platform, enables its customers to interact with their clients from smartphones or smart watches. SalesForce itself uses a help system where anyone in the company can ask the human resources department a question, and anyone in that department can respond in real-time from anywhere in the world, including during dinner.

Beyond creating bite-sized work that can be accomplished outside of a traditional workplace, to better serve these employees, we will need to develop a more thorough understanding of their entire work system and develop new methods for measuring work in now atypical environments. The widespread adoption of smart devices highlights how we have replaced a former institution of work and leisure: the publishing of physical newspapers and other media. With a monumental amount of information available from the device that fits in your hand, many people find the appeal of printed material greatly diminished, and the future of the medium is hotly debated.

## TREND 3: RESILIENCE

In 2001, the healthcare industry was greatly impacted by an Institute of Medicine Report "To Err Is Human." The report catalyzed a long process of hospital safety improvement that started with rules and procedures for everything. Similarly, the airline industry was built on having a strict hierarchy of command between pilot and copilot. Both industries are adopting a more synergistic and flexible model called a *resilient systems* approach (Jentsch, Barnett, Bowers, & Salas, 1999). The core precept in resilience is that you cannot predict every problem and therefore must support people in the system by giving them some flexibility and autonomy to work synergistically.

For a more detailed description of resilience, refer to *Resilient Healthcare* (Hollnagel et al., 2013), *Aviation: Architecting Resilient Systems: Accident Avoidance and Survival and Recovery from Disruptions* (Jackson, 2009), *Energy Resilient Buildings & Communities: A Practical Guide* (Levite & Rakow, 2015), and *Resilience Engineering in Practice: A Guidebook* (Pariès, 2013). In the next decade, resilience will find its way into other industries too.

## TREND 4: AFFECTIVE SYSTEMS DESIGN

Affective design, hedonomics, aesthetic engineering, emotional design, and kansei engineering are all synonyms (on a macro scale) for a field that seeks to enrich user experience by creating products and systems that draw the user in. Affective design uses quantitative measurement and testing of user experience or selection to create devices that are demonstrably more desirable or enjoyable to use. Affective device studies have been carried out in domains such as blood glucose meters, mobile phones, and websites (Nathan-Roberts & Liu, 2012).

To illustrate, many people laud Apple for the beauty of their iPhone, but technophiles will point to devices with higher sound quality, larger storage, smaller envelopes, fewer software restrictions, etc., that did not have the cult following or transformational power of the popular smartphone. However, through a combination of design and business decisions, the iPhone in all its permutations remains massively popular. Similarly, the Herman Miller Aeron chair has been awarded many design awards, including Business Week's "Design of the Decade" award and an induction into the Museum of Modern Art's Hall of Fame (Herman-Miller, 2014), and has become an icon of human factors, and users often refer to how comfortable it is (PRNewswire, 1999). However, research has shown that although users rated the chair more desirable and aesthetically pleasing than other high-end ergonomic office chairs, the postural angles and pressure zones experienced by users are worse than with other ergonomic chairs (Nathan-Roberts, Chen, Gscheidle, & Rempel, 2008).

Affective design has been gaining devotees for the last 20 years, but it is largely focused on individual products. With the increase in macroergonomic thinking in our field, affective design will transition to affective systems (Nathan-Roberts & Yin, 2015). An example (albeit unhealthy) affective system is a casino, in which the ornate physical objects, reward structure, layout, and customer service are all designed together to encourage gambling (Nathan-Roberts & Brennan, 2013).

A more positive example can be found in the documentary *Stitch in Time* (Canadian Broadcasting Corporation, 1985), which illustrates the beneficial use of affective design in Shouldice Hospital's care. In this case, aspects of both treatment and business processes have been streamlined to provide the most effective surgical treatment for patients with hernias.

## TREND 5: CULTURAL HUMAN FACTORS

When a patient is being discharged from the hospital with clear instructions to avoid spicy food, how do you think their cultural background might affect their compliance with the treatment plan? The definition of "spicy" might mean something different for someone who primarily eats Latin American or southeast Asian cuisine versus, say, traditional Scandinavian foods. Cultural background is similarly relevant as the digital revolution has given us a new power to communicate and work across the globe. Increasingly, "international business" will become "business," and our daily work will involve a greater array of interactions and practices.

Cultural human factors seeks to understand where these relationships may occur and better design the system to ameliorate these types of problems. Cultural human factors has been predominately researched in terms of patient–provider trust (Montague, 2010) and health information technology (Valdez, 2010). For a more complete description of cultural human factors, we recommend *Cultural Ergonomics: Theory, Methods, and Applications* (Smith-Jackson, Resnick, & Johnson, 2013).

## TREND 6: MIXED METHODS

Human factors stemmed from the time savings and reduction of musculoskeletal disorders found by Fredrick Taylor and Frank and Lillian Gilbreth in the late 1800s and early 1900s (Frederick W. Taylor, *Man as a Mechanism in the Factory*). Concerned primarily in efficiency gains in repetitive tasks, Taylor and the Gilbreths were instrumental in generating methods of quantitative human performance data (Pollard, 1974). The field of human factors has come a long way in the last 125 years, with work increasingly focused on understanding more nuanced ramifications to changes. With the goal of understanding these subtle interactions, mixed-methods research is growing in popularity.

The goal behind mixed-methods research is to strengthen quantitative findings with corresponding qualitative feedback and narratives to provider a larger contextual view of the findings. Mixed methods is especially useful in complex operational environments where a larger view of the system is needed. Mixed methods is different from sociotechnical work systems because it is a specific set of methods that either have a quantitative (numerical data) bent with some qualitative (nonnumerical data) methods used in conjunction (quantitatively driven approaches and designs), qualitative methodological focus with supporting quantitative analysis (qualitatively driven approaches and designs), or a balanced approach in which quantitative and qualitative research methods share the stage equally to build a rich picture relying on both methodologies equally (interactive or equal-status designs). Conversely, sociotechnical systems are a framework for viewing problems, not a specific set of methods. For a more complete explanation of mixed-methods research, we direct you to *Designing and Conducting Mixed Methods Research* (Creswell & Plano Clark, 2007) and the *NIH Best Practices for Mixed Methods* report (Creswell, Klassen, Plano Clark, & Smith, 2011).

## CONCLUDING REMARKS

Human factors is an engineering science and discipline in part because the theories, methods, and tools that are used to improve a system in one domain will have applicability outside of that domain. Here, domain refers to an area of industry or a type of tool (aviation, automotive, web, mobile, healthcare, etc.). Domains can also be subdivided into very particular domains (laparoscopic surgery versus endoscopic surgery).

Domains currently rich in human factors research include: aging, automotive, aviation, education, energy, healthcare, human–computer interaction, human performance modeling, occupational health, safety, and virtual environments. Leverage the learnings, methods, and tools from other domains. For example, crew resource management, originally from the aviation field, is also useful in surgery and improving outcomes in other teams where there is a strong hierarchical component. If you are working to improve vigilance in security screening tasks, for example, look first at what has been done in your domain and then look across domains at vigilance in general. The healthcare and aviation domains both have a history of work in vigilance (Cruz, Boquet, Detwiler, & Nesthus, 2003; Slagle & Weinger, 2009). The findings of many domains can be integrated to widespread effect. It is in this spirit of collaboration and perseverance that great advances can and will be made in the study of human factors.

As technology rapidly changes, so does our ability to design and streamline our processes to make them more effective and efficient. Cross-domain development enables our greater understanding of human factors and, thus, better use of the technology around us and of that yet to come. Our success depends not only on research but also on the use of that knowledge across the wide range of human needs and goals.

## REFERENCES

Cruz, C., Boquet, A., Detwiler, C., & Nesthus, T. (2003). Clockwise and counterclockwise rotating shifts: Effects on vigilance and performance. *Aviation, Space, and Environmental Medicine, 74*(6), 606–614.

Canadian Broadcasting Corporation. (1985). *Stitch in time* [video documentary]. Toronto, Canada: Canadian Broadcasting Corporation.

*Frederick W. Taylor: Man as a Mechanism in the Factory.* Retrieved from http://www.teachspace .org/personal/research/management/taylor.html

Herman-Miller. (2014). *The Herman Miller Aeron chair is voted one of the 12 best designs of the past 100 years.* Retrieved from http://www.hermanmiller.com/content/dam/hermanmiller /page_assets/about_us/news_events_media/apac/HMAeron1Of12BestDesigns.pdf

Hibbard, J. H., Mahoney, E. R., Stockard, J., & Tusler, M. (2005). Development and testing of a short form of the Patient Activation Measure. *Health Services Research, 40*(6p1), 1918–1930.

Holden, R. K., Carayon, P., Gurses, A. P., Honnakker, P., Hundt, A. S., Ozok, A. A., & Rivera-Rodriguez, A. J. (2013). SEIPS 2.0: A human factors framework for studying and improving the work of healthcare professionals and patients. *Ergonomics, 56*(11), 1669–1686.

Hollnagel, E., Braithwaite, J., & Wears, R. L. (2013). *Resilient healthcare.* Boca Raton, FL: CRC Press.

Jackson, S. (2009). *Aviation: Architecting resilient systems: Accident avoidance and survival and recovery from disruptions.* Hoboken, NJ: Wiley.

Jentsch, F., Barnett, J., Bowers, C. A., & Salas, E. (1999). Who is flying this plane anyway? What mishaps tell us about crew member role assignment and air crew situation awareness. *Human Factors: The Journal of the Human Factors and Ergonomics Society, 41*(1), 1–14.

Levite, B., & Rakow, A. (2015). *Energy resilient buildings and communities: A practical guide.* Lilburn, GA: Fairmont Press.

Montague, E. N. H. (2010). Trust in medical technology by patients and healthcare providers in obstetric work systems. *Behaviour and Information Technology, 29*(5), 541–555.

Nathan-Roberts, D., & Brennan, P. F. (2013). Patient Affective System Design; Informatics Aspects of Engaging Care. Poster. *AMIA 2013 Annual Symposium.* Washington, DC.

Nathan-Roberts, D., Chen, B., Gscheidle, G., & Rempel, D. (2008). Comparisons of seated postures between office tasks. *Proceedings of HFES 52nd Annual Meeting.* New York. Doi: 10.1177/154193120805200902

Nathan-Roberts, D., & Liu, Y. (2012). Comparison of design preferences for mobile phones and blood glucose meters. *Proceedings of HFES 56th Annual Meeting.* Boston, MA.

Nathan-Roberts, D., & Yin, S. (2015). Affective health design; Hospital service systems improvement case study. *Proceedings of 6th International Conference on Applied Human Factors and Ergonomics and Affiliated Conferences.* Las Vegas, NV.

Pariès, J., Wreathall, J., Hollnagel, E., & Woods, D. (2013). *Resilience engineering in practice: A guidebook.* Boca Raton, FL: CRC Press.

Pollard, H. R. (1974). *Developments in management thought.* New York: Crane, Russak, & Company, Inc.

PRNewswire. (1999). *Herman Miller's Aeron(R) work chair named among ELITE 'Designs of the Decade'.* Retrieved from http://www.prnewswire.com/news-releases/herman-millers-aeronr-work-chair-named-among-elite-designs-of-the-decade-77276082.html

Slagle, J. M., & Weinger, M. B. (2009). Effects of intraoperative reading on vigilance and workload during anesthesia care in an academic medical center. *Anesthesiology, 110*(2), 275–283.

Smith-Jackson, T. L., Resnick, M. L., & Johnson, K. T. (2013). *Cultural ergonomics: Theory, methods, and applications.* Boca Raton, FL: CRC Press.

Valdez, R. S. (2010). Designing culturally-informed consumer health IT: An exploration and proposed integration of contrasting methodological perspectives. *Proceedings of the Human Factors and Ergonomics Society 54th Annual Meeting.* San Francisco, CA.

Vincente, K. (2004). *The human factor: Revolutionizing the way people live with technology.* London, UK: Routledge.

Wickens, C. D., Lee, J. D., Liu, Y., & Gordon-Becker, S. (2003). *Introduction to human factors engineering* (2nd ed.). London, UK: Pearson.

## APPENDIX 10-A: USEFUL TIPS FOR THE HUMAN FACTORS PRACTITIONER

Dan Nathan-Roberts and David Schuster

1. Do not blame the user. Remember that they are doing their best within the larger context in which they live (competing goals, work stress, family stress, etc.).
2. Use methods from other domains to test in your domain and to devise successful solutions.
3. Design out the possibility of failure. Training is not the solution; neither is labeling. If you cannot physically make the dangerous choice, you will not need a warning label. To paraphrase this concept: "If you cannot design, train; if you cannot train, select; if you cannot select, warn."
4. Know when you should get help. (If you do not know, then you probably need help.)
5. Remember that you are not the average user. When devising a solution, have others try it, play with it, and break it. Utilize the input of diverse users, including across generations (the infamous "grandma tester"). Do this early and often.
6. Accelerate the failure cycle. If you are not failing early and frequently, you are not testing things that matter. Iterations lead to better solutions.
7. Do not make checklists, instructions, or surveys too long; chances are, no one will read the entire thing.
8. Dare to be boring. Yes, conventions can be arbitrary. That the left-side hotel sink handle usually controls hot water and the right-side controls cold water is not because it is better that way, but it capitalizes on people's expectancies.

## APPENDIX 10-B: SUPPLEMENTAL RESOURCES FOR THE HUMAN FACTORS PRACTITIONER

Kelli Sum, Danielle Ishak, Michael Cataldo, Kelsey Hollenback, and Dan Nathan-Roberts

### LEISURE READING

*Set Phasers on Stun: And Other True Tales of Design, Technology, and Human Error* (2 Sub Edition) by Steven Casey
Book Review by Kelsey Hollenback

In *Set Phasers on Stun*, Casey presents 20 real-world cautionary tales of the mismatch between machines and the humans who are meant to operate them, with dangerous and often tragic consequences; the second edition has been updated to include two new chapters and a prologue in which Casey sets out his purpose in writing the book. As the author states, his choice to tell stories instead of collect case studies is intended to strip away theoretical discussion and root-cause analysis to focus on what he contends is the real point: that, when technology is designed without considering human limitations, failure is certain to result.

While the book's narrative style is easily accessible, the writing is clunky at times, and his use of the third person omniscient can edge into the realm of speculative fiction (e.g., when he describes the thoughts and feelings of a World War II navy diver who maintained radio silence during his mission and died in the course of completing it). However, he also conveys deep sympathy for the human actors; *Set Phasers on Stun* is frequently an uncomfortable read, much in the way that watching a train wreck in slow motion is uncomfortable. Casey's great strength is in inviting the reader to step away from blaming "human error" and toward a broader systems view until the reader reaches the conclusion that, as one of the book's unfortunate subjects, when the difference between life and death is one push of a button, Captain Kirk eventually *will* forget to set the phasers on stun.

### *The Atomic Chef: And Other True Tales of Design, Technology, and Human Error* by Steven Casey
Book Review by Michael Cataldo

Casey's follow-up to *Phasers* encompasses the importance of human-centered design in today's evolving society. As the author states, "the discipline of ergonomics, or human factors engineering, seeks to address human characteristics, capabilities, and limitations and reflect them in the design of the things we create to make them easier to use, more reliable, and safer." This book demonstrates past failures, some historic and horrifying, and contains a total of 20 stories that reveal what could best be called "design-induced errors."

The stories described herein illustrate the troubles *users* have faced and, consequently, why catastrophic accidents have occurred. For example, in the story "A Kid in a Car," a child fatally suffers from a design-induced error of a window switch. This story is a valuable lesson for all: that anybody can fall victim to these errors in any environment.

Although this collection of stories can scare the reader, they serve as examples of why we must consider all plausible factors in a design. For readers who are interested in human factors, this book is highly suggested for its broad scope on design and technology. It is with this quality that the stories take readers through airports, amusement parks, hospitals, battlefields, and even ships and spacecraft. *The Atomic Chef* lays a great foundation for discussion on the meaning of design and why it is so crucial to our society.

### *The Human Factor: Revolutionizing the Way People Live With Technology* by Kim Vicente
Book Review by Kelli Sum

For readers with little to no knowledge of human factors but an initial interest, *The Human Factor* by Kim Vicente is an accessible book with which to explore these concepts. The book highlights the disparity between the development of technology and the usability for humans. Vicente suggests that technology is present in five key areas of society: political, organizational, team, psychological, and physical. With this framework is carried the expectation that technology will enhance and improve these areas of life; accidents and fatalities, it purports, occur time after time again due to a lack of consideration or future planning of how the human will be using the technology. This book may influence the reader to view mistakes and accidents in a different way. Instead of placing the blame on the operator, the reader may begin to see a pattern of poorly designed systems or products being the root cause of past incidents, which provides valuable insight to novice and experienced readers alike.

## GENERAL TEXTBOOKS

*Introduction to Human Factors Engineering* (2nd Edition) by Christopher D. Wickens, John D. Lee, Yili Liu, and Sallie Gordon-Becker
Book Review by Danielle Ishak

Providing foundational knowledge of the field of human factors and its many topics, *An Introduction to Human Factors Engineering* is a great overall resource. The book is not a leisurely read, as some of the concepts are complex and technical; however, it is written simply to make the concepts easier to comprehend. Basic examples and problems are provided to aid the reader when learning the process of human factors computations related to biomechanics and cognition.

Some of the topics in this book include introductions to research methods, visual sensory systems, cognition, decision-making, displays, controls, work physiology, human–computer interaction, safety, and more. When one is reading about the different topics, the book provides just enough foundational information about the topic as to provide a good understanding of the depth of the topics.

This book is excellent for a classroom setting or for an introduction to human factors so that readers may get a broader technical understanding of all of the different topics in the field. It can also serve as a reference book for human factors academics or practitioners who work in one of the many topics in the field and desire a quick technical sourcebook.

*Human Factors in Simple and Complex Systems* by Robert W. Proctor and Trisha Van Zandt
Book Review by Kelli Sum

Proctor and Van Zandt have provided in *Human Factors* a textbook that provides readers with an academic introduction to human factors. With this book, the reader can expect to gain an understanding of the historical foundation of human factors, perceptual factors, cognitive factors, action factors, and environmental factors and their applications. In addition, the reader will learn about the connection between conceptual and empirical foundations of the field. The majority of the chapters explain past studies; however, periodically boxes are placed throughout the text, providing additional information about how a concept is applied in a real-world situation.

The target audience is undergraduate- and graduate-level university students in introductory courses of human factors or applied cognition. Mathematical and science equations are not explained in great depth, which makes this textbook an accessible option for readers with a less-technical background. On the other hand, readers with previous knowledge of science or mathematics will be able to grasp equations quicker but may struggle with research methods and theory. Therefore, this textbook would be an optimal choice for a variety of people who are initially interested in human factors and want an overall understanding of its theory and application.

## METHODOLOGICALLY SPECIFIC OR DOMAIN-SPECIFIC BOOKS

Bridger, R. S. (2008) *Introduction to ergonomics* (3rd ed.). CRC Press: Boca Raton, FL.

Carayon, P. (2011). *Handbook of human factors and ergonomics in healthcare and patient safety* (2nd ed.). Boca Raton, FL: CRC Press.

Chaffin, D. B., Anderson, G. B. J., & Martin, B. J. (2006). *Occupational biomechanics* (4th ed.). New York: Wiley.

Eastman Kodak Company. (2003). *Kodak's ergonomic design for people at work* (2nd ed.). New York: Wiley.

Freivalds, A., & Niebel, B. (2013). *Niebel's methods, standards, & work design*. New York: McGraw-Hill.

Groover, M. P. (2006). *Work systems: The methods, measurement & management of work*. Upper Saddle River, NJ: Pearson.

Sanders, M. S., & McCormick, E. J. (1993). *Human factors in engineering and design* (7th ed.). New York: McGraw-Hill.

Stanton, N., Hedge, A., Brookhuis, K., & Salas, E. (Eds.). (2004). *Handbook of human factors and ergonomics methods*. Boca Raton, FL: CRC Press.

## SHORT COURSES AND OTHER OPPORTUNITIES FOR ABRIDGED FURTHER EDUCATION

University of Michigan Short Course (http://umich.edu/~driving/shortcourse/): This course, held annually in Ann Arbor, Michigan, spans two weeks and covers the design of systems, products, and services to make them easier, safer, and more effective for human use. The coursework emphasizes principles and concepts and includes application examples. It is best for human factors professionals and those new to the field.

CQPI SEIPS course (http://cqpi.wisc.edu/SEIPS-short-course.htm): Based in Madison, Wisconsin, the SEIPTS Human Factors and Patient Safety annual short course is designed to help students understand human factors and system engineering, as well as how this approach to patient safety can improve system performance, prevent harm, and aid in system recovery. It is best for healthcare professionals, quality-improvement specialists, management personnel, patient safety offices, and all others interested in human factors engineering and patient safety.

AAMI HFE Medical Device Approval Course: These courses span several lengths from a single day to several and review the HE75 medical device guidance.

COECH UC Berkeley Lectures and Workshops: These short courses span various lengths and address commonly used concepts and theories in the field.

Human Factors Education by Entertainment (https://sites.google.com/site/educationbyentertainment/): Held by Dr. Ronald G. Shapiro, these light-hearted activities are held at various events and add a fun component to serious human factors and design topics.

Embry-Riddle Aeronautical University Center for Aerospace Safety/Security Education (CASE) (http://proed.erau.edu/): A full array of courses present opportunities for learning in aviation-related fields.

HFES ErgoX (http://www.hfes.org/web/HFESMeetings/ergoX.html): This three-day conference, targeted toward industry professionals, is held around the United States and focuses on emerging and relevant topics in the field of human factors.

Harvard T.H. Chan School of Public Health's Ergonomics and Human Factors: Strategic Solutions for Workplace Safety and Health (https://ccpe.sph.harvard.edu /programs.cfm?CSID=EHF1015): Held on the East Coast of the United States, this four-day conference provides healthcare and occupational health professionals the opportunity to learn about and investigate solutions for workplace safety and health issues.

Centre Quebecois de Formation Aeronautique Pilot Selection Course (in French; http://www.cqfa.ca/pro/index.php?id=98): Over two days in Chicago, Illinois, this course helps air carriers and flight academies develop and improve methods of selection and recruitment of pilots.

Centre Quebecois de Formation Aeronautique Pilot Interviewing Workshop (in French; http://www.cqfa.ca/pro/index.php?id=1020): In this one-day course held in Chicago, Illinois, participants use a blended format of tutorials presentations and practical workshops to develop, administer, and evaluate an interview involving several interviewers

National Ergonomics Conference & ErgoExpo (www.ergoexpo.com): This expo welcomes participants over three days in the United States for an industry-wide symposium on human factors, featuring lectures, vendors, and networking opportunities.

Online Education in Human Factors and Ergonomics: A number of colleges and universities provide remote-learning opportunities in human factors. The Human Factors and Ergonomics Society maintains a complete list of programs that currently includes Auburn University; California State University; Dominguez Hills; California State University, Northridge; Centre Quebecois de Formation Aeronautique; Florida Institute of Technology; North Carolina State University; Pennsylvania State University; Purdue University; University of Alabama, Huntsville; University at Buffalo, State University of New York; University of Idaho; University of Maryland; and the University of Virginia.

Additional continuing education resources can be found on the Human Factors and Ergonomics Society website (http://www.hfes.org/Web/EducationalResources /educresourcesmain.html).

# Index

Page numbers followed by f and t indicate figures and tables, respectively.